Agricultural Drainage Engineering

ABOUT AUTHOR

Prof. Hiteshkumar V. Parmar did his B. Tech. (Agril. Engg.) in 2001 and M. Tech. (Agril. Engg.- Soil and Water Engineering) in 2003 from College of Agricultural Engineering and Technology, Gujarat Agricultural University, Junagadh, Gujarat. He specializes in Soil and Water Engineering with emphasis on Agricultural Drainage Engineering, Watershed Management, Hydrology, Water Harvesting, Fluid Mechanics, Remote Sensing and GIS, etc.

He is working as Assistant Professor at the College of Agricultural Engineering and Technology, JAU, Junagadh since year 2008. He has been an active teacher, researcher and has guided number of B. Tech. (Agricultural Engineering) students for their research project in his area of specialization. He also worked as Project Officer (2003-2005) on Watershed Management, Natural Resource Management with Foundation for Ecological Security (Earlier, National Tree Grover's Co-operative Federation- NTGCF), Dahod, a sister concern of National Dairy Development Board (NDDB) - Anand.

He passed GATE in year 2001 and NET for Lectureship/ Assistant professorship in the professional subject Soil and Water Engineering conducted by ASRB, ICAR, New Delhi in year 2006. He also qualified written examination of Indian Forest Service, 2007.

He has many research publications in national and international journals and conferences. He is a life member of Indian Society of Agricultural Engineers (ISAE). He is the secretary of ISAE-Gujarat chapter. He was the member secretary of the committee for the preparation of Comprehensive District Agriculture Plan (C-DAP) of Junagadh district.

AGRICULTURAL DRAINAGE ENGINEERING

Hiteshkumar V. Parmar

SCIENTIFIC
PUBLISHERS (INDIA)

Published by:

Scientific Publishers (India)
5 A, New Pali Road, P.O. Box 91
Jodhpur 342 001 (India)

Branch Office
Scientific Publishers (India)
4806/24, Ansari Road, Daryaganj
New Delhi - 110 002 (India)

E-mail: info@scientificpub.com
Website: www.scientificpub.com

ISBN:9788172338596

9 788172 338596
Price: 1950/-

IS

Typeset by Rajesh Ojha
Printed in India

PREFACE

Agricultural drainage system is planned to remove excess water from the crop land, whether coming through irrigation system, rainfall or runoff, before it creates waterlogging or high salinity condition. The sustainability of irrigated agriculture in many areas has been affected to a great extent by the necessity for drainage. Lack of adequate drainage was the primary cause of failure of certain agricultural civilizations. Starting as an art, born out of sheer necessity two centuries ago, drainage engineering has now acquired a scientific basis. In many regions of the world the provision of drainage facility in agricultural lands tends to become necessary with the passage of time. Without a suitable drainage network, excess precipitation tends to stagnate in large quantities for too long a period as well as during the critical crop growth periods, thereby severely decreasing the yield of crops.

The book on "Agricultural Drainage Engineering" deals with the problem of drainage, drainage investigation, salt balance, surface drainage systems, subsurface drainage systems, reclamation of degraded land, designs and installation of drainage system, drainage materials, fabrication and testing of tiles, gypsum requirement, design of envelope, special drainage systems, supplementary protection structures, classification of soil and water, cost estimation, operation and maintenance of drainage systems, etc. The best effort has been made to acquaint the students about the practical application of agricultural drainage engineering like measurement of hydraulic conductivity in field and in laboratory, preparation of iso-bath and isobar maps, installation of piezometers and observation wells, design and installation of drainages systems, etc. The book is profusely illustrated with sketches and is not bulky.

I believe that book "Agricultural Drainage Engineering" will assist in imparting better skills and understanding to Under Graduate (UG) and Post Graduate (PG) students, research scholars and professionals in the field of Agricultural Engineering (Irrigation, Drainage, Soil and Water Conservation Engineering, Watershed Management, etc.), Agriculture (Soil Science, Soil Chemistry, Soil and Water Management, Natural Resources Management, etc.), Civil Engineering (Water Resource Management, Water Supply and Drainage Engineering, etc.), etc. In addition to its use as university level text and reference, students of polytechnics in agriculture engineering and agriculture, the book will be helpful to the students as a reference for preparation of competitive examinations like GATE/JRF/SRF/NET/ARS, etc. Suggestions for improving the book are always welcome and will be incorporated in the next edition.

Hiteshkumar V. Parmar

LIST OF TABLES

Table No.		Page No.
1.1	Tolerance levels of some of the major crops	11
2.1	Soil Texture - Hydraulic Conductivity Relationship	27
3.1	Hydraulic Conductivity of Different Soil Types	43
3.2	Observation table for the Auger-Hole Method	52
3.3	Observation table for the Inversed Auger-Hole Method	56
3.4	Observation table for the Inversed Auger-Hole Method	62
3.5	Observation table for Falling Head Permeameter	65
3.6	Observation table for Constant Head Permeameter	67
4.1	Comparison between Interceptor Drain and Relief Drain	70
4.2	Runoff Coefficient "C" for Agricultural Watersheds (Soil Group B)	81
4.3	Hydrologic Soil Group Conversion Factors	81
4.4	Runoff Coefficients for the Rational Formula versus Hydrologic Soil Group (A, B, C, D) and Slope Range	82
4.5	Distribution of 6-hr. rain within 6 hours	85
4.6	Hydrologic Soil Cover Complex	85
4.7	Antecedent Moisture Content (AMC)	86
4.8	Hydrologic condition	86
4.9	Curve Number Values for Different Land Use and Cover Conditions	86
4.10	Conversion of curve numbers from AMC-II to AMC-I and AMC-III	87
4.11	Value of Manning's n	89
4.12	The Permissible Velocities in Open Drains	89
4.13	Recommended Side Slopes for Drainage Channel	89
5.1	Average Depth and Spacing of Tile Drains (adopted from Schwab *et al.;* 1993)	95
5.2	Equivalent Depth 'd_e' for Commonly Used Pipe Drains (for drains with diameter = 80 to 200 mm)	104
5.3	Drainage Coefficient for Tile Drains in Humid Regions	117
5.4	Minimum Grades for Pipe Drains	118
5.5	Values for Manning's Coefficient 'n' for Pipe Material	118
6.1	Gradation relationships between soil and diameters of graded granular envelope material	144

Table No.		Page No.
10.1	Classification of Salt Affected soils	183
10.2	Chemical Properties of various Amendments for Reclaiming Alkali	187
10.3	Gypsum Requirement and Exchangeable Sodium	188
10.4	Gypsum Requirement Using Pure Gypsum, One-to-one Replacement of Sodium by Calcium and Complete Dissolution and Mixing of Gypsum	189
10.5	Observation table for Gypsum requirement	191
10.6	Crop Tolerance to Alkalinity of Soil	192
10.7	Average soil salinity tolerated by specific crop	196
11.1	Water Quality Ratings (CSSRI, Karnal, 1972)	205
11.2	Relative Salt Tolerance of Crops	206
11.3	Soil Classification as per pH and EC	207
11.4	Classification of Soil by EC and SAR	208
11.5	Soil Salinity Classes and Crop Growth	209
11.6	Salt concentration and salinity classes	209
11.7	Drainage Water Quality Criteria for Irrigation Purposes	212
11.8	Effect of diluted drainage water on Wheat Yield	212
11.9	Plants and Irrigation water salinity	213
12.1	Water balance factors	227

LIST OF FIGURES

Fig. no.		Page no.
1.1	Types of Drainage System Only	18
1.2	Combined Drainage System	19
2.1	Observation Well	32
2.2	Construction of Water Table Contour Lines by Linear Interpolation	40
2.3	Water Table Contour Lines	41
2.4	Water Table Contour Map	41
3.1	Auger-hole method	51
3.2	Measurement for the Auger-Hole Method	51
3.3	Infiltration process beneath a Cylinder Infiltrometer	53
3.4	Measurements for the Inversed Auger-Hole Method	54
3.5	Fall of the water level, recorded with the inversed auger hole method, plotted against time (a).	56
3.6	Fall of the water level, recorded with the inversed auger hole method, plotted against time (b).	56
3.7	Infiltration form a water filled auger hole into the soil (Inversed auger hole method)	57
3.8	Measurements for the inversed auger hole method	57
3.9	Fall of the water level, recorded with the inversed auger-hole method, plotted against time (the symbol indicating the slope of the line should be α)	58
3.10	Hanging Column Apparatus for Laboratory Determination of Drainable Porosity	61
3.11	Falling Head Permeameter	64
3.12	Constant Head Permeameter	66
4.1	Levelling	75
4.2	Layout for Lateral, Branches, Sub-Main, Main	76
4.3	Profile for drain	77
5.1	Steady Flow into a Ditch	98
5.2	Definition sketch for Hooghoudt's equation	100
5.3	Illustration of the parameters involved in Hooghoudt's equation for Steady state condition	101
5.4	Actual and idealized patter of flow towards a subsurface drain	103
5.5	Equivalent depth for the water conducting layers below the drain	104
5.6	Illustration of the distance parameters involved in the drainage of sloping lands, with entrance resistance taken into account	105
5.7	The shape of the water table obtained with the finite element method as published by Fipps and Skaggs (1989)	107

Fig. no.		Page no.
5.8	Steady flow into Drains in Layered Media	108
5.9	Nomograph for the solution of Hooghoudt's drain spacing formula	109
5.10	Definition sketch for USBR equation	110
5.11	Depiction of symbols for solution of USBR equation	111
5.12	Nomograph for equivalent depth	112
6.1	Three Edge Bearing Test	136
6.2	Sand Bearing Test	137
6.3	Coconut Fiber Envelope	141
6.4	Example of the Design of a Grave Envelope	143
7.1	Drainage Layout	148
7.2	Random Drainage System	148
7.3	Herringbone Type Drainage System	149
7.4	Gridiron Drainage System	149
7.5	Interceptor Type Drainage System	150
7.6	Downstream location of the drain is marked by Pegging	151
7.7	Stakes are placed for levelling	151
7.8	Tools to install drains by hand	153
7.9	Subsurface Drainage Pipe	156
7.10	Subsurface Drainage Section	157
8.1	Mole Plough and Mole Channel	165
8.2	Flow of Water toward the Well	168
9.1	Blind inlet	172
9.2	Surface inlet	172
9.3	Gravity outlet	173
9.4	Outlets	174
9.5	Silt basin	176
9.6	Junction	178
9.7	Manhole	179
9.8	Drain Bridge	179
10.1	Relationship between Gypsum requirement and Soil pH	189
12.1	The concept of 4 reservoirs with hydrological inflow and outflow components	214
12.2	Illustration of a surface water balance during periods of high rainfall	220
12.3	Month wise Water balance	221
12.4	Reuse of drainage water pumped with wells for irrigation	221
12.5	Runoff and discharge with time	222
12.6	The salt concentration of the percolation water depends on the degree of mixing of infiltration water with the soil aggregates.	223
12.7	Examples of the relation between hydrology and salinization.	225
12.8	Example of salinization in uneven land	226

CONTENTS

1	**INTRODUCTION**		**1-21**
	1.1	Irrigation and Drainage Development	1
	1.2	Drainage Problems in India	2
	1.3	Waterlogging	4
	1.4	Classification of Waterlogging Areas	5
	1.5	Causes of Waterlogging	5
	1.6	Control of Waterlogging	8
	1.7	Agricultural Land Drainage	9
	1.8	Objectives of Drainage System	10
	1.9	Effect of Poor Drainage on Soil and Plants	11
	1.10	Benefits of Drainage	12
	1.11	Agricultural Drainage Systems	14
	1.12	Types of Drainage System	17
	1.13	Comparison of Surface and Subsurface Systems	20
2	**DRAINAGE INVESTIGATION**		**22-42**
	2.1	Basic Data Required	22
	2.2	Collection of Existing Data	23
	2.3	Types of Investigations	23
	2.4	Soil survey	25
	2.5	Water Sources Survey	29
	2.6	Ground Water Investigations	29
	2.7	Piezometers and Observation Wells	30
	2.8	Observations	33
	2.9	Methods for Measuring Water Levels	35
	2.10	Frequency of Observations	36
	2.11	Analysis of Data	36
	2.12	Groundwater Maps	39
3	**HYDRAULIC CONDUCTIVITY**		**43-69**
	3.1	Hydraulic Conductivity	43
	3.2	Hydraulic Conductivity Variability	44

	3.3	Methods of Determination of Hydraulic conductivity	46
	3.4	The auger-hole method (small-scale in-situ method)	48
	3.5	Inversed auger-hole method	57
	3.6	Large-scale in-situ methods	58
	3.7	Drainable Porosity	59
	3.8	Measurement of hydraulic conductivity falling head permeameter apparatus	63
	3.9	Determination of the coefficient of permeability using constant head permeameter.	65
4	**SURFACE DRAINAGE SYSTEM AND DESIGN**		**70-90**
	4.1	Area Which Requires Surface Drainage	70
	4.2	Land Formation for Surface Drainage	71
	4.3	Surface Drainage Systems	73
	4.4	Surface Drains	74
	4.5	Leveling for the Drains	75
	4.6	Design of Surface Drainage System	77
	4.7	Hydrologic Design of Open Drain	78
	4.8	Rational Method of Predicting Peak Runoff Rate	80
	4.9	Estimation of Runoff using Curve Number Technique	83
	4.10.	Hydraulic Design of Open Drain	87
5	**SUBSURFACE DRAINAGE SYSTEM AND DESIGN**		**91-124**
	5.1	Benefits of Sub-Surface Drainage	91
	5.2	Methods of Subsurface Drainage	93
	5.3	Tile Drain System	94
	5.4	Deep Open Drains	94
	5.5	Depth and Spacing of Tile Drains:	94
	5.6	Steady State Condition	96
	5.7	Unsteady State Condition	110
	5.8	Drainage Coefficient	114
	5.9	Hydraulic Design of Subsurface Drainage Systems/ Tile Drains	116
	5.10	Economics and Cost Analysis of Drainage System	119
6	**DRAINAGE MATERIALS**		**125-145**
	6.1	Drain Pipes	125
	6.2	Characteristics of Tiles	126
	6.3	Clay Tiles	126
	6.4	Concrete Tiles	127
	6.5	Plastic Drain Pipes	128

	6.6	Pipe Accessories	130
	6.7	Testing of Drainage Tiles	131
	6.8	Mechanical Characteristics	134
	6.9	Testing the Strength of Concrete or Clay Tiles	136
	6.10	Possible Future Test Methods	137
	6.11	Envelope for subsurface drainage system	139
	6.12	Envelope Material	140
	6.13	Design Criteria of Envelope	142
	6.14	Design of Drainage Envelope	143
7	**INSTALLATION AND MAINTENANCE OF DRAINAGE SYSTEMS**		**146-163**
	7.1	Layout of Surface and Subsurface Drainage Systems	147
	7.2	Construction Methods	150
	7.3	Machinery for Drain Pipe Installation	151
	7.4	Installation of Subsurface Drainage System	152
	7.5	Planning and Preparatory Aspects for Drainage Installation	154
	7.6	Supervision and Inspection during Installation of Drainage System	157
	7.7	Monitoring	157
	7.8	Maintenance of Drainage System	159
8	**SPECIAL DRAINAGE SYSTEMS**		**164-170**
	8.1	Mole Drainage System	164
	8.2	Stability of Mole Channel	165
	8.3	Factors Affecting Mole Channel Stability	165
	8.4	Vertical Drainage	167
	8.5	Suitability of Vertical Drainage	169
	8.6	Bio-drainage	169
9	**SUPPLEMENTARY PROTECTION STRUCTURES**		**171-180**
	9.1	Inlets	171
	9.2	Outlets	172
	9.3	Outlet Protection Structures	175
	9.4	Silt Basins	176
	9.5	Breathers	177
	9.6	Relief Pipes	177
	9.7	Junction Boxes	177
	9.8	Manholes	178
	9.9	Drain Bridges	179
	9.10	Gradient Reducers	180

10	**RECLAMATION OF SALT AFFECTED SOILS**	**181-201**
10.1	Salt Affected Soils	181
10.2	Sources of Salinity	183
10.3	Causes of Water-logging	185
10.4	Source of Salts in Water	185
10.5	Reclamation of Alkali Soils	185
10.6	Reclamation of Saline Soil	193
10.7	Reclamation of Saline - Alkali Soils	199
11	**SOIL AND WATER QUALITY**	**202-213**
11.1	Water Quality	202
11.2	Analysis of Wate	203
11.3	Classification of Water	205
11.4	Salt Tolerant Crops	206
11.5	Classification of the Soil	207
11.6	Determination of Chemical Properties of Soil and Water	209
11.7	Conjunctive use	211
11.8	Cyclic use	213
12	**WATER AND SALT BALANCE OF AGRICULTURAL LAND**	**214-229**
12.1	Water Balance in Agriculture Land	214
12.2	Water Balance of the Surface Reservoir	216
12.3	Water Balance in the Root Zone	216
12.4	Water Balance in the Transition Zone	216
12.5	Water Balance in the Aquifer	217
12.6	Combined Water Balances	217
12.7	Steady State Condition	219
12.8	Examples of Water Balances	220
12.9	Salt Balances in the Agricultural Land	222
12.10	Examples of Salt Balances in Agriculture Land	224
13	**GLOSSARY**	**230-234**
	REFERENCES	**235-238**

CHAPTER – 1

INTRODUCTION

With 3,287,590 square kilometers, India is the seventh largest country in the world and with 944.58 million inhabitants the second most populated country. India has five principal physio-graphic regions: The Himalayan Mountains, the Indo-Gangetic plains, the Great Indian Desert, the Deccan Plateau and the coastal mountain belts.

Almost the entire territory is situated in the tropical or subtropical region. The climatic features are diverse on both regional and seasonal basis due to its size, peninsularity, topography and geographical conditions. The climate ranges from tropical wet to arid and semi-arid, and weather conditions have even greater variations. The climatic conditions, which are characterized by frequent droughts and floods, influence to a great extent India's utilization of water resources. About 80 per cent of annual precipitation is concentrated in a period of about 3 months, during the monsoon season. The main water resources in India are rainfall and snowmelt of glaciers in the Himalayas that feed the rivers.

Floods and droughts are regular phenomena. Floods occur between June and October as a result of cyclonic storms. An area of 40 million hectares (or 12 per cent of the total land area) is estimated to be vulnerable to floods. These areas are situated predominantly in the eastern part of the Ganges basin and the north-eastern part of the country.

1.1 IRRIGATION AND DRAINAGE DEVELOPMENT

India's irrigated agriculture has been fundamental for economic development and poverty alleviation. According to the World Bank Sectoral Report (1998), 28 per cent of India's GDP and 67 per cent of its employment is based on agriculture, and about two-thirds of the total agricultural output on irrigated agriculture. In 1997, the irrigated area comprised 90 million hectares. About 80 per cent of India's farmland grows main foods such as grains and pulses, and the major cultivated cereals are millet, maize, rice and wheat. Groundwater contributes 35 per cent to irrigation, and is particularly used in minor schemes; surface water supplies 52 per cent and is primarily used in major and medium schemes, and 13 per cent in minor schemes. In 1991, the total drained area was estimated at 5.8 million hectares, which is 12 per cent of the irrigated area (FAO 1999).

Irrigation can be dated back to prehistoric times. Old irrigation structures still exist in different parts of the country. Modern irrigation started under the British colonial regime in the nineteenth century with the construction of canals, storage facilities and river diversion. In 1947, the year of India's independence, an area of about 22 million hectares were under irrigation. Since then, a major emphasis of India's governments has been the implementation of irrigation projects with the aim to increase food production.

However, during rapid irrigation development, planning and construction of drainage facilities in irrigated areas did not keep pace. Investment in drainage has been widely neglected compared with irrigation. After implementation and operation of many irrigation projects, a rise in the water table with following degradation of soil through waterlogging and salinity was observed. The first notice of waterlogging in irrigated fields was made in 1850 in the State of Punjab followed by other irrigation and canal projects in other States (CWC 1997).

Land drainage development with subsurface drains is of recent origin comparing with India's long tradition of surface drainage. Vertical drainage has been applied on a limited scale. In addition to structural measures, bio-drainage and various crop-water management techniques have been applied to deal with drainage problems (CWC 1997).

In 1974/75, the Indian government launched the centrally financed program for Command Area Development. Its primary objective has been to increase efficient utilization of irrigation water and to improve agricultural productivity in irrigation commands. This program is realized through major and medium irrigation projects comprising different components of water resource management. Concerning drainage, the program includes on-farm development through field channels and field drains, the reclamation of waterlogged areas (since 1996) and the encouragement of participatory irrigation management. Furthermore, it contains development and maintenance of main and intermediate drainage systems. The program is expected to be implemented by Command Area Development Authorities (CADA) under the Ministry of Water Resources. Where CADAs have not been constituted, the program is administered through the States' Irrigation Departments (MOWR 2000).

A new generation of irrigation projects – the Water Resource Consolidation Projects - is assisted by the World Bank. Their main objectives are the improvement of institutional and technical capability of managing water resources, rehabilitation and completion of irrigation schemes and farmers' participation.

1.2 DRAINAGE PROBLEMS IN INDIA

In earlier times, drainage problems were handled with moderate canal realignment and improvement of drainage systems, but today's problems are much more complex. The large scale development of irrigation projects caused waterlogging and soil salinity in many irrigation commands. Despite the experience gained in surface drainage so far, maintenance and control of weeds are still inadequate. Kitamura *et al.* (1997) report that attention for drainage development and improvement focused more on irrigation-induced problems in arid and semi-arid regions, and less on rainfall-induced problems. Where investment for drainage has been made, lack of proper maintenance has led to siltation of many drains. Irrigation and drainage infrastructure is generally in a poor state. Rehabilitation requirements represent an increasing part of the investment.

Estimations about the total area affected by waterlogging and soil salinity are of a very questionable nature with respect to their magnitude and relevance. The existing lack of standardization for classifying the problems of waterlogging and salinity is mentioned as a reason Singh and Datta (1997). Kitamura *et al.* (1997) note the difficulty to differentiate between rainfall- induced and irrigation-induced waterlogging. Data on the extent and distribution of salinity face similar difficulties because it needs to be differentiated between areas affected by secondary, i.e. irrigation-induced, salinity and primary salinity.

This results in different figures for the magnitude of the problem. The Central Water Commission estimates the area affected by waterlogging and soil salinity as follows:

- About 8.5 million hectares are waterlogged, whereof about 2.46 million hectares are suffering from waterlogging under irrigation.
- A total of 5.5 million hectares are affected by salinity, whereof about 3.06 million hectares are estimated to be affected due to irrigation-related problems.
- The total affected land under waterlogging and salinity is 14 million hectares of which 5.52 million hectares would suffer from inadequate drainage in irrigated lands (CWC 1997).

India's wide range of climatic, physiographic and geo-hydrological conditions are leading to varying patterns of waterlogging and salinity. The Central Water Commission (1997) differentiates drainage in five main geographical zones:

1. North-West India

This zone is located in the States of Punjab, Haryana, north-western Rajasthan and western Uttar Pradesh and has semi-arid to arid climate. Due to the lack of natural outlets, excess irrigation water and salt cannot be evacuated efficiently, leading to the twin-problem of waterlogging and salinity.

2. Central Peninsular India

This zone, covering the States of Madhya Pradesh, Maharashtra, Karnataka and western Andhra Pradesh, is of a hard rock area generally with shallow soil cover and undulating topography. Drainage problems are caused by seepage from canals and percolation from irrigated fields and are of a local nature. Waterlogging and salinity occurs in some areas where the soils are more mineralized.

3. Eastern Plains and Deltas

This zone comprises the States of eastern Uttar Pradesh, Bihar and West-Bengal, Orissa, Andhra Pradesh and Tamil Nadu. Its main problem is waterlogging due to high monsoon rainfall. Persistently high water levels in the main drains lock the outfall. In some areas introduction of irrigation has aggravated waterlogging. Soil salinity represents no problem because of high rainfall.

4. Coastal Area of Western India

In the State of Gujarat, large areas of the coastal zone are affected by excess salinity, which is primarily of marine origin. Irrigation leads to an aggravation of the problem. Additional factors that increase the problem are deficient salt leaching due to low rainfall, the poor natural soil drainability and rising water tables.

5. Sodic Land of Western Gangetic Plains

Sodic soils are predominantly found in the States of Haryana, Punjab, Uttar Pradesh and parts of Bihar, mostly in the upper layers of depressions. It is considered that sodicity occurs due to prolonged ponding of surface water. At some places where groundwater is used for irrigation, which contains carbonates, soil sodicity occurs.

1.3 WATERLOGGING

Waterlogging is the accumulation of excess water in the root zone of the soil. Part of the water that infiltrates into the soil will be stored in the soil pores and will be used by the crop; another part of the water will be lost as deep percolation. When the percolating water reaches that part of the soil which is saturated with water, it will cause the water table to rise. If the water table reaches the root zone, the plants may suffer. The soil has become waterlogged. Drainage is needed to remove the excess water and stop the rise of the water table. UN Food and Agricultural Organization (FAO/UNESCO, 1973) defined waterlogged areas as those soils are temporarily saturated or where the groundwater table is too shallow such that capillary rise of water encroaches upon the crop root zone and may even reach the soil surface. National Commission of Agriculture, Govt. of India (NCA, 1976) has defined an area as waterlogged when the watertable causes saturation of crop root zone soil, resulting in restriction to air circulation, decline in oxygen and increase in carbon dioxide levels. The productivity of land may get affected when the root zone of the plants gets flooded with water, and thus become ill aerated as all the air pockets available in the soil arc filled by with water molecules and it disturbs the suitable ratio of soil, water and air and ultimately stops exchange of gases. Oxygen present in the air is not only needed by human being but is also needed by plants. The life of the plants depends upon the nutrients like nitrates. The form in which nitrate is consumed by the plants is produced by the bacteria, under a process called nitrification. These bacteria need oxygen for their survival. The supply of oxygen gets cutoff when the lands becomes waterlogged, resulting the death of these bacteria and fall in production of plants food (i.e. nitrates) and consequent reduction in plants growth, and hence, crop yields are reduced.

Even if irrigation water is of very good quality, it will contain some salts. So, bringing irrigation water to a field also means bringing salts to that field. The irrigation water is used by the crop or evaporates directly from the soil. The salts, however, are left behind. This process is called salinization. If these salts accumulate in the soil, they will hamper the growth of crops. Salinization is the accumulation of soluble salts at the soil surface, or at some point below the soil surface, to levels that have negative effects on plant growth and/or on soils.

1.4 CLASSIFICATION OF WATERLOGGING AREAS

- *Waterlogged areas :* Watertable within 2 m from the land Surface
- *Potential area for waterlogging :* Watertable between 2-3 m from the land surface
- *Safe area :* Water cable below 3 m from the land surface

The following are the common approaches to express the watertable depth below the soil surface:

- Pre monsoon (April/May) depth to watertable.
- Post-monsoon (October/November) depth to watertable.
- Seasonal (monsoon/winter/summer) or annual average depth to watertable.
- Sum of the number of days when watertable is shallower than a specified depth

Norms for the classification of different categories of waterlogged areas in India:

Waterlogged areas

(i) Watertable within 2 m from soil surface during pre-monsoon (April/May),

(ii) Watertable within 1m from soil surface during post-monsoon (October/November).

Critical area: When the watertable is between 2-3 metres from the soil surface during pre-monsoon and/or between 1-2 metres during post-monsoon, it is considered as critical.

Potential area: In monsoon, irrigated areas with watertable between 3-5 metres during pre-monsoon may be considered as potential areas for waterlogging.

1.5 CAUSES OF WATERLOGGING

Waterlogging may happen due to various factors as enumerated below:

1. *Over irrigation:* When policy of intensive irrigation is adapted the maximum irrigable area of a certain region is irrigated. This may lead to too much of irrigation, resulting in more percolation and subsequent but slow rise in watertable. For this reason, to avoid waterlogging, the policy of extensive irrigation (i.e. irrigation spread over wider region) should suppress the policy of intensive irrigation;
2. *Seepage:* Water from adjoining high lands may seep into the sub-soil of the affected land and may raise the watertable, sometimes occur water stagnation in low lying areas. Water may seep through the bed and sides of the adjoining canals, reservoirs, etc situated at a higher level than the affected land, resulting in buildup of high watertable;
3. *Hard pan:* Water seepage below the soil moves horizontally (i.e. laterally) but may find an impervious obstruction causing the rise of watertable on

the upstream side of the obstruction. Similarly an impervious stratum may occur below the top layers of pervious soils. Is such cases water seeping through the pervious soils will not be able to go deep, and thus quickly resulting in high watertable.

4. *Lack of outlets:* Soils having less permeable sub stratum (such as clay) below the top layers of pervious soils, will not able to drain the water deep into ground, and hence resulting in high water Level in affected soil;
5. *Inadequate surface drainage:* Excessive storm water falling over the land and excess irrigation water should be removed and should not be allowed to percolate below. If proper drainage is not provided the water will constantly percolate and will raise the level of underground reservoir,
6. *Excessive rains:* Excessive rainfall may create temporary waterlogging and in absence of good drainage, it may lead to contribute watertable built-up and continued waterlogging;
7. *Flood:* If land is continuously submerged by floods, water loving plants like grass, weeds etc may grow which obstruct the natural Surface drainage and thus increasing the chances of temporary waterlogging;
8. *Irregular topography: On steep terrain, the water is drained out quickly. On flat* or irregular terrain having depressions, the drainage is very poor. AH the factors lead to greater detention of water on land and hence more percolation and increased watertable*
9. *Climatic Factors:* Erratic rainfall distribution within a short wet season followed by a long dry season is the specific features of a monsoon climate. Heavy monsoon rain causes accumulation of runoff. Its ill effects are more severe in flat terrains and low-lying areas, even during a relatively smaller rainfall in some of the non-monsoon months. In the absence of drainage, natural or man-made, the accumulated water gets more opportunity time to percolate down and causes the water table to rise. In heavy textured soils or black cotton soils, due to excess wetness of the soil, delay in the sowing of winter crop after the withdrawal of monsoon is a common feature in HI drained soils, which reduces the crop yield. High evaporation and evapotxanspiration in the dry season increases the salt concentration on the soil surface and beneath the soil (in the root zone).
10. *Lack of Catchment Area Protection:* Deforestation, overgrazing, unplanned urbanization and faulty cultivation practices cause soil loss due to erosion. Waterlogging over the downstream lands, siltation and pollution of downstream natural streams reducing the capacity of man-made surface drains and reservoirs, and help to deteriorate the water quality. The intensity of the flood hazard increases if the carrying capacity of natural streams has reduced due to siltation.
11. *Adverse Topography:* Are as with adverse topography are known by different local names such as *Bhal* (as flat as the forehead) lands in Gujarat, *Diara* (concave) and, *Tal and Cbaur* (flood plains of low elevation) lands in Bihar and Uttar Pradesh. There are low-lying lands

with severe restriction of natural drainage in several parts of India Examples are the large areas in the Indo-Gangetic plain, the delta regions of the seas and the below-sea level cooperative rice farms in the Kuttanad region of Kerala. In many of such low lands or land with adverse topography, the monsoon cultivation season is lost due to high level of land inundation following incessant heavy rains, floodwater accumulation and backwater flow.

12. *Canal Seepage:* Seepage from unlined or lined but ill-maintained canal in the irrigated areas causes long duration wet and sometimes marshy condition in the adjoining agricultural lands ultimately results in rise in water table and development of salinity and alkali conditions. This is a major problem in irrigated agriculture due to which, not only the valuable irrigation water is lost but also the water that is lost causes other problems like salinity and alkalinity.

13. *Injudicious use of Irrigation Water:* Field to field irrigation and over-irrigation due to uncertainty in the release of canal water both in time and in quantity, are the examples of injudicious use of irrigation water. Besides, unconsolidated lands, inadequate land preparation, mostly ungated and many unauthorized outlets and an absence of scientific rotation of canal water lead to poor irrigation efficiency. These, in turn, cause temporary surface waterlogging and watertable rise.

14. *Other Man-made Causes:* Unplanned construction of roads, rail lines and railway embankments and under-capacity culverts lead to restriction in the natural flow of water, results in waterlogging. Adoption of unscientific cropping practices such as growing more water requiring crops against the planned cropping pattern in a canal irrigated area results in applying more than the anticipated quantity of water over relatively smaller areas. This causes excess water problem in certain parts and water deficit over the other parts.

15. *Poor Water Quality:* Brackish ground and surface water, when used in irrigation for want of adequate quantity of fresh water, adds undesirable salts to the agricultural lands. Disposal of untreated municipal and industrial effluent on the land and into the natural streams causes serious surface and groundwater quality problems.

16. *Inherent Problem with the Soil:* High salt content in the soil due to properties of the parent rock material, poor water transmission characteristics in heavy soils and high evapotranspirative water demand in a major part of the year lead to salinity and alkali problems, this is more conspicuous in the semi-arid regions.

17. *Lack of Proper Prioritization:*Land development and reclamation, construction of field drains and drain maintenance, are not considered as priority items of work in an irrigation command area. The working funds for such works are always limited. Many of these works are planned but are not executed adequately and in time This leads to the development of waterlogging and salt problems, which become expensive to tackle at a later stage.

18. *Lack of Monitoring:* Monitoring of the gradual changes in the health of the land and the water resources are seldom done routinely in any command area after the introduction of irrigation. Problems of drainage (watertable rise) and salinity become known when it is too late and the remedy becomes expensive,
19. *People's Participation:* Absence of a participatory approach by the irrigation water users to manage the water resource at the farm level and developing suitable management alternatives to tackle the upcoming problems leads to gradual degradation of the physical and the chemical environment of the agricultural land.
20. *Lack of Policy Guideline:* The Ministry of Water Resources, Government of India, has brought out a National Water Policy in 1987, which was subsequently revised in 2002. This policy has recognized water as a scarce resource and considers its planning and management as matters of utmost urgency for its optimal, economical and equitable use. On the other hand, the National Land Use and Wastelands Development Council under the Chairmanship of the Prime Minister, has adopted, in Toto, a National Land Use Policy Outline drafted and submitted by the National Land Use and Conservation Board in 1986. The Council's meeting also stipulated 19 action points to be pursued by various State Land Use Boards constituted mostly under the Chairmanship of the Chief Ministers. Unfortunately, the progress on taking action on the stipulated points has not been encouraging, presumably because the National Land Use Policy Outline was not a document formally endorsed by the Parlia-ment However, all the Union Ministries related to use and management, including the Ministry of Water Resources and ICAR were among the parries when this policy outline was adopted. There has been little interactive development in taking cognizance of various lines and directives given in this Policy Outline.

1.6 CONTROL OF WATERLOGGING

It is evident that waterlogging can be controlled, only if the quantity of water into the soil below is checked and reduced. To achieve this, the inflow of water into the underground reservoir should be reduced, and the outflow from this reservoir should be increased. The various measures adopted for controlling water logging enumerated are as below:

1. *Lining of canals and watercourse:* Attempt should be made to reduce the seepage of water from the canals and watercourses. This can be achieved by lining them. It is very effective method to control waterlogging.
2. *Reducing the intensity of irrigation:* In areas where there is a possibility of waterlogging, intensity of irrigation should be considerably reduced. Only a small portion of irrigable land should receive canal water in one particular season. The remaining areas can receive water in the next season.

3. *By introducing crop rotation:* Certain crops require more water and other require less water. If a field is always sown with a crop requiring more water, the chances of waterlogging are more. In order to avoid this, a high water requiring crop should be followed by the one requiring less water and then by one almost no water. Rice may be followed by wheat, and wheat may be followed by a dry crop like cotton.
4. *Optimum use of water:* It has been said earlier that only a certain amount of water gives best results. Less than that or more than that reduces the yield. But most of our cultivators are unaware of these techniques, and they feel that by using more water they can increase their yields. Therefore, they try to use more and more water. This can be checked by educating the cultivators through proper propaganda. Moreover, revenue should not be charged on the basis of irrigated area, but should charged on the basis of quantity of water utilized, A strict watch should be kept at the outlet in order to Stop undue tapping.
5. *Providing intercepting drains:* Intercepting drains along the canals should be constructed, wherever necessary, and the constructed drains should be maintained properly. These drains can prevent the sub-soil water from reaching the area likely to be waterlogged
6. *Provision of an efficient drainage system;* An efficient drainage system should be provided in order to drain away the storm water and the excess irrigation water. A good drainage system consists of surface drains, sub-surface drains as well as interceptor drains.
7. *Improving the natural drainage of the area:* To reduce percolation, it is necessary that water should not be allowed to stand for a longer period. Some relief in this direction can be obtained by removing the obstruction in the path of natural flow. This can be achieved by removing bushes, rubbles, etc. and maintaining the proper slopes of the natural drainage lines.
8. *Introduction of lift irrigation:* Lift irrigation utilizes the underground water for irrigation. It, therefore, helps in lowering the water tables through tube-wells. Canal irrigation may be substituted by lift irrigation, in areas, which are likely to be waterlogged.

1.7 AGRICULTURAL LAND DRAINAGE

The sustainability of irrigated agriculture in many areas has been affected to a great extent by the necessity for drainage. Lack of adequate drainage was the primary cause of failure of certain agricultural civilizations. Starting as an art, born out of sheer necessity two centuries ago, drainage engineering has now acquired a scientific basis. Drainage of agricultural lands has benefited from advances in soil physics and hydrology which have led to improved methods of design in this field. Developments of new materials for pipes and machinery used for constructing of surface ditches and underground piping have led to impressive progress in drainage systems. Integrated design of irrigation water supply system and drainage water collection and conveyance system is highly desirable. Just as the blood vessel system consisting of arteries and veins is essential for the survival and health of animals and humans, the survival and health of irrigated agriculture, especially in humid and clayey areas depends on the provision of integrated irrigation and drainage components. Avoidance of water-logging, creation of well aerated root zone, and removal of objectionable salts contribute to increased crop yields on a long term basis.

Need for Drainage

Need for a drainage system, vis-a-vis, irrigated lands can best be appreciated by considering the problem-situations, the objectives to be fulfilled by drainage schemes, the design requirements and components of an efficient drainage system. In many regions of the world the provision of drainage facility in agricultural lands tends to become necessary with the passage of time. Without a suitable drainage network, excess precipitation tends to stagnate in large quantities for too long a period as well as during the critical crop growth periods, thereby severely decreasing the yield of crops. Even excess irrigation supplies permitted without due care can lead to waterlogging that would be detrimental to the crops. In such situations the soil-moisture content remains higher than the field capacity for quite a long time - especially in humid climates with nearly steady and heavy rainfall spells and in flat lands with fine textured (clayey) soils, excessive soil moisture would affect crop yield in the absence of adequate drainage. This problem is more common in areas wherein average annual evaporation rates do not exceed average annual rainfall rates.

It is well known that there are soils that contain objectionable salts in their profile; also the irrigation water supplied itself may bring in objectionable amounts of salts. Under these circumstances there is a strong possibility of such salts getting accumulated on the soil surface and in the top root-zone as a result of root uptake, which seriously affects the crop growth and its yield. Such salt-accumulation should be periodically got rid off. Many irrigation systems that flourished in the historical past (along with the civilizations that they nurtured) faced their doom due to the salinity problem. Thus, to ensure sustainability, an irrigation system needs to be complemented with a drainage system. *Leaching* is the common method adopted for getting rid of salts. The leached water (also called 'leachate') should be safely lead away from the cultivated lands. Thus, drainage systems are intended to deal with those issues of excess quantity of water (surface and subsurface), and of objectionable salts in the root zone.

1.8 OBJECTIVES OF DRAINAGE SYSTEM

The objectives of a drainage system may be summarized as follows:

(a) Crop Growth and Nutrition

To ensure favorable root zone in sufficient depth, which is well aerated and with sufficient moisture and nutrients. Further, a good drainage system helps reducing the washing away of seeds and fertilizers by flood waters. A well developed root system, active over a long crop period leads to high yields.

(b) Quality of Land

(i) Waterlogged lands can be reclaimed through a proper drainage system. The asset-value of the land thus increases.

(ii) Poorly drained area will suffer from increase in the incidence of malaria; thus, the provision of drainage leads to improved sanitation.

(iii) A proper control of moisture in the soil mantle avoids soil baking and cracking.

(c) Salinity Control

(i) To keep the water table at an optimum depth so as to prevent salt effervescence at the surface and in the top soil.

(ii) To help safely dispose off the highly saline leachate while leaching is being undertaken.

Table 1.1. Tolerance levels of some of the major crops

Highly tolerant (up to 10 g/l)	Moderately tolerant (up to 5 g/l)	Sensitive (up to 2.5 g/l)
Date palm	Wheat Tomato Oats Alfalfa Rice Maize Flax	Red clover Peas Beans
Barley	Potato Carrot Onion	Sugar cane
Sugar beet	Cucumber Pomegranate Fig	Pear Apple Orange
Asparagus	Olive	Prune Plum Almond
Spinach	Grape	Apricot Peach

1.9 EFFECT OF POOR DRAINAGE ON SOIL AND PLANTS

1. Water fills the soil pores and displaces air and obstructs the gases, which are given off by the roots;
2. The soils become blocked and inefficient for the exchange between gases in the soil and in the atmosphere;
3. The oxygen content in the soil is less;
4. Extremely slow rate of diffusion of gases is observed through such soils;
5. In waterlogged soil, gas exchange is confined to a fraction of the top inch of soil, below this oxygen is non-existent;
6. After the dissolved oxygen in waterlogged soils has been consumed, anaerobic decomposition of organic matter takes place. This results in the production of reduced organic compounds, such as methane or marsh gas, methyl compounds, etc;
7. Because of the decline in rate of decomposition, nitrogen tends to remain locked up in the organic residues;
8. The decline in the rate of transpiration, the decay of roots and lack of formation of new roots takes place;

9. Adverse effect on the uptake of nutrients by plants;
10. Health hazard is created by increase in population of mosquitoes, which breed in ponds and small puddles in the field.

1.10 BENEFITS OF DRAINAGE

1. If the fields are properly graded and outlets and ditches for the disposal of water are provided, this situation will be remedied and more healthful and pleasant environments for human habitation will be created.
2. The crops can root more deeply. Drainage removes excess water and helps the roots to grow into the increased soil volume. Consequently, more nutrient becomes available to the plant and its anchorage with soil improves to support against lodging under wet soil and high wind conditions.
3. Optimum soil temperature is needed for healthy germination of seeds. A high water table results in a soil that does not warm up readily in the spring. Germination of crops is delayed and the seed may infect rot before it germinates. The drained soils become warmed soil.
4. Due to hot summer temperature, water that is ponded because of irrigation or rainfall, kills planted grasses or legumes. Drainage with proper grading reduces the damage due to scalding (burning due to hot water) increases the yield.
5. Drainage creates congenial root environment, which is essential for the growth of beneficial soil bacteria, which convert soil organic matter and fertilizer to available plant food.
6. The unfavorable root environment exists in the areas of high watertable. Plant diseases are more active under these conditions. Fungus growth is particularly prevalent.
7. In wet and heavy soils, field operations are not only difficult but also destroy the soil structure.
8. Drainage brings the field to the field capacity (working condition) sooner and keeps so for a longer time. Field operations become easier and less time and labour consuming.
9. Harmful salts are washed away and leached out of the root zone by the drainage system, maintaining a favourable salt-water balance in the agricultural land.
10. Removal of excess water by drainage reduces the specific heat of the soil-water system. A warmer and more favourable soil temperature regime extends over a longer period. This accelerates bacterial activity and plant growth.
11. Plant availability of Nitrogen of the applied fertilizer increases in well-drained soils and loss of Nitrogen due to denitrification is minimized. If the soil is heavy, saline-sodic and waterlogged, the positively charged ammonium ions of the fertilizer fail to adsorb on the negatively charged soil clay complex in the presence of stronger cations and ammonia

volatilization loss occurs. Some ammonia, however, get lost through the leachate of the subsurface drainage system during the initial period of reclamation of heavy and salt affected soils. But once reclaimed, such losses are completely avoided.

12. The choice of crops is greater. Drainage is not crop-specific and a good variety of crops can be grown successfully in well-drained agricultural lands.
13. Drainage helps early planting due to improved soil-water regime and as a result longer effective photoperiod becomes available to the plant for its growth.
14. Drainage is essential for salinity-alkalinity control in arid/semi-arid regions.
15. Waterlogged and ill-drained soils are the breeding grounds of agents that cause diseases to men and livestock such as malaria, dengue, foot rot and liver fluke. Mosquitoes breed profusely in stagnated water bodies and affect the livestock and human beings. Drainage removes these problems.
16. Aeration in root zone: One of the benefits of installing a drainage system to remove excess water is that the soil is better aerated. This leads to a higher productivity of crop land or grassland because:
17. There will be fewer weeds.
18. Fertilizers will be used more efficiently.
19. There will be less denitrification.
20. The grass swards will be better.

Other benefits of well-drained soils are

1. The land is more easily accessible.
2. The land has a greater bearing capacity.
3. The soil has a better workability and tilth.
4. The period in which tillage operations can take place is longer.
5. The activity of micro-fauna (e.g. earthworms) is increased, which improves permeability.
6. The soil structure is better, which also improves permeability.
7. Soil temperatures are higher, so that crops (particularly horticultural crops) and grasses can be grown earlier.
8. When drainage makes it possible to control the water table, the benefits that follow are:
9. The root zone cannot become salinized by the capillary rise of saline groundwater.
10. Leaching is made possible.
11. In its turn, the benefits of leaching are:
12. It prevents increases in soil salinity in the root zone, thus making irrigated land use sustainable in the long term.

13. By removing salts, it allows salt-sensitive crops, or a wider range of crops, to be grown.
14. It makes it possible to reclaim salt-affected soils, thus bringing new land into cultivation.

1.11 AGRICULTURAL DRAINAGE SYSTEMS

Agricultural drainage systems are systems that make it easier for water to flow from the land, so that agriculture can benefit from the subsequently reduced water levels. The systems can be made to ease the flow of water over the soil surface or through the underground, which leads to a distinction between "surface drainage systems" and "subsurface drainage systems". Both types of systems need an internal or "field drainage system", which lowers the water level in the field, and an external or "main drainage system", which transports the water to the outlet.

A surface drainage system is applied when the water logging occurs on the soil surface, whereas a subsurface drainage system is applied when the water logging occurs in the soil. Although subsurface drainage systems are sometimes installed to reduce surface water logging and vice versa, this practice is not recommended. Under certain conditions, however, combined surface/subsurface drainage systems are quite feasible.

Agricultural drainage systems do not necessarily lead to increased peak discharges. Although this may occur, especially with surface drainage, the reduced water logging can lead to an increase in the temporary storage of water on or in the soil during periods of peak rainfall, so that peak discharges are indeed reduced. A drainage engineer should see to it that the flow of water from the soil occurs as steadily as possible instead of suddenly. In irrigated, submerged rice fields, a form of temporary drainage is required whereby the drainage system is only allowed during the harvest period. If allowed to function continuously, excessive quantities of water would be lost. Such a system is therefore called a checked drainage system. More usually, however, the drainage system should function as regularly as possible to prevent undue water logging at any time.

The above definition of agricultural drainage systems excludes drainage systems for cities, highways, sports fields, and other non-agricultural purposes. Further, it excludes natural drainage systems. Agricultural drainage systems are artificial and are only installed when the natural drainage is insufficient for a satisfactory form of agriculture.

Important drainage methods are mentioned below:

1. Surface drainage, including land grading and smoothing
2. Subsurface drainage or pipe drainage
3. Vertical (Tube well) drainage
4. Bio-drainage

Factors that are of prime importance in drainage are enumerated below:

1 Topographic factors

2. Soil factors

3. Water and climatic factors
4. Crop factors

Design Requirements of Drainage System

The two main design requirements of the system are given below:

(i) The carrying capacity should be adequate to safely pass the design storm runoff. The outfall of the system should be adequate to freely receive this discharge and should not lead to backing up.

(ii) Cross-section of the pipe lines should not suffer from scouring or silting. The surface of the pipes should inhibit weed growth. The initial expenditure and the maintenance expenditure should be cost effective.

An agricultural drainage system is typically made up of following components:

(a) Drainage water collection (Field system)

A field drainage system, which prevents ponding water on the field and/or controls the water table. The field drainage system is a network that gathers the excess water from the land by means of field drains, possibly supplemented by measures to promote the flow of water to these drains. A field drainage system can be a surface drainage system (to remove excess water from the surface of the land) or a subsurface drainage system (to control the water table in the soil). In surface drainage, field drains are shallow graded channels, usually with relatively flat side slopes. In subsurface drainage, field drains can be either open drains or pipe drains. Open drains and pipe drains have the same function. The difference between them is the way they are constructed: an open drain is an excavated ditch with an exposed water table; a pipe drain is a buried pipe

(i) Overland flow and surface drains,

(ii) Shallow drainage, including mole drain and shallow ditches, and

(iii) Groundwater drainage, with drain pipes, or deep ditches.

(b) Drain water conduction (main system)

A main drainage system, which conveys the water away from the farm. The main drainage system is a water-conveyance system that receives water from the field drainage systems, surface runoff and groundwater flow, and transports it to the outlet point. The main drainage system consists of some collector drains and a main drainage canal. A collector drain collects water from the field drains and carries it to the main drain for disposal. Collector drains can be either open drains or pipe drains. The main drain is the principal drain of an area. It receives water from collector drains, diversion drains, or interceptor drains (= drains intercepting surface flow or groundwater flow from outside the area), and conveys this water to an outlet for disposal outside the area. The main drain is often a canalized stream (i.e. an improved natural stream), which runs through the lowest parts of the agricultural area.

(i) Tertiary, secondary and primary ditches,

(ii) Tertiary, secondary and primary pipe drains.

(c) Outlet

Terminal point at a natural open water body such as river, lake or sea, reached either by gravity or assisted by pumping.

Outlet of a Field Drainage System

The first and most important consideration in good drainage is the outlet. Irrespective of the type of drainage system that will be installed on the farm, a good outlet is a prerequisite for success. If there is no way to evacuate the water away from the field, the drainage system will not work.

The water level at the outlet defines the drainage base. It determines the hydraulic head available for drainage flow. The drainage base is different for different points in a drainage area. For the field drainage system, the drainage base is the water level in the collector drains, whether they are pipes or open drains. For the collector drainage system, the drainage base is the water level in the main drain. And for the main drainage system, it is the water level below the gravity outlet structure during critical periods for crop growth, or the minimum water level at the pumping station.

Care should be taken to ensure that the water level in the recipient drain, whether it is a collector or the main drain, is below the required water level in the field, especially in the period when drainage is most important (e.g. in the rainy season).

Drainage Requirement

For the design of a drainage system, the drainage requirement or the drainable surplus has to be known. This is the amount of water that must be removed from an area within a certain period so as to avoid an unacceptable rise in the levels of the groundwater or surface water. Removing the drainable surplus has two advantages:

- It prevents waterlogging by artificially keeping the water table sufficiently deep.
- It removes enough water from the root zone so that any salts brought in by irrigation cannot reach a concentration that would be harmful to crops.

Calculating the drainage requirement is a major problem in many irrigated areas. The natural conditions in these areas are diverse, and different water resources may be involved in the calculations. Therefore field work has to be carried out to find out what the general features of the groundwater regime are, and the water and salt regimes and their balances have to be studied. A proper understanding of these regimes allows the drainage engineer to predict how they will be affected by drainage.

To calculate the drainage requirement, an analysis has to be made of the overall water balance of the study area. Water balances are often assessed for an average year. Waterlogging and salinity problems, however, are not of the same duration or frequency every year. Therefore there is often a need to assess water balances, not only for an average year, but also for specific years (e.g. a very dry year or a year with extreme rainfall), or even for specific periods (e.g. the growing season or the irrigation season).

Dissolved Salts in the Groundwater

All groundwater contains salts in solution. The type of salts depends on the geological environment, the source of the groundwater, and its movement. Irrigation is also a source of the salts in the groundwater. It not only adds salts to the soil, but also dissolves salts in the root zone. Water that has passed through the root zone of irrigated land usually contains salt concentrations several times higher than that of the originally applied irrigation water. Evapotranspiration tends to concentrate the salts at the surface of the land (Figure 29), but when they are dissolved, they increase the salinity of the groundwater. Therefore, highly saline groundwater is often found in arid regions with poor natural drainage.

1.12. TYPES OF DRAINAGE SYSTEM

Natural drains are those formed in nature as depicted in the geomorphology of an area. Topographically, the valley line between ridges forms a natural drainage. Wherever feasible, these should be used to drain waterlogged marshy lands and offending ponds. Artificial drains in agricultural land are engineered works meant to effectively improve the soil moisture regime of the cultivated land. From the functional view point, artificial drains may be classified as follows:

(a) Surface drains

(b) Subsurface drains, and

(c) Combination of surface and subsurface drains.

There is a recent development wherein the system can serve as a drain for part of the period and as an irrigation source during other period. These are referred to as Combined Drainage and Sub-Irrigation (CDSI) systems. Figure depicts these various possibilities.

Surface Drainage only

(A) Drainage only

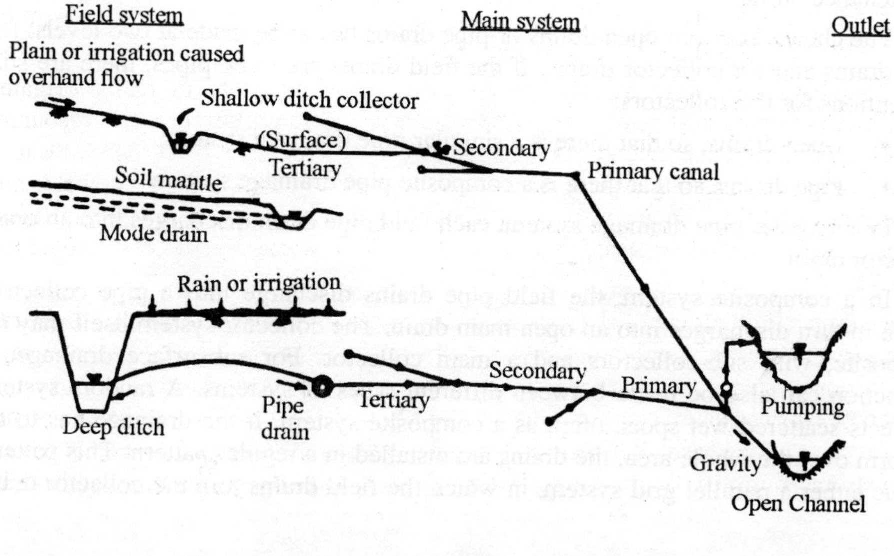

Eco drain

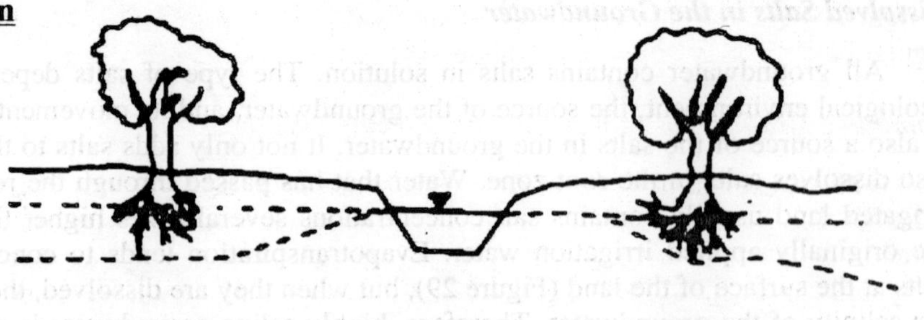

(B) Combined Drain-cum-Sub Irrigation System (CDSI System)

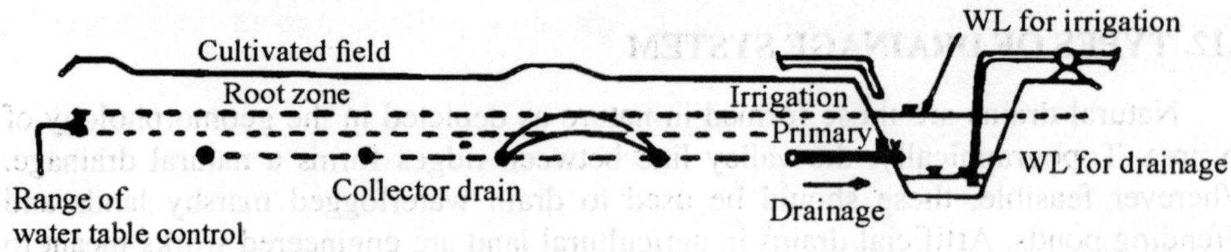

Figure 1.1. Types of Drainage System Only

Subsurface drainage systems

A subsurface drainage system is a system for the removal of excess water and dissolved salts from the soil, using the groundwater as a "vehicle". A subsurface drainage system is a man-made system that induces excess water and dissolved salts to flow through the soil to pipes or open drains, from where it can be evacuated

If it is decided to install a subsurface drainage system, a choice has to be made between open drains or pipe drains. Open drains have the advantage that they can receive overland flow and can thus also serve as surface drainage. The disadvantages are the loss of land, the interference with the irrigation system, the splitting up of the land into small farm blocks, which hampers farming operations, and that they are a maintenance burden.

The choice between open drains or pipe drains has to be made at two levels: for field drains and for collector drains. If the field drains are to be pipes, there are still two options for the collectors:

- Open drains, so that there is a singular pipe drainage system;
- Pipe drains, so that there is a composite pipe drainage system.

In a singular pipe drainage system, each field pipe drain discharges into an open collector drain.

In a composite system, the field pipe drains discharge into a pipe collector, which in turn discharges into an open main drain. The collector system itself may be composite, with sub-collectors and a main collector. For subsurface drainage, a distinction can also be made between different types of systems. A random system connects scattered wet spots, often as a composite system. If the drainage has to be uniform over the whole area, the drains are installed in a regular pattern. This pattern can be either a parallel grid system, in which the field drains join the collector drain at right angles, or a herringbone system, in which they join at sharp angles. Both regular patterns may occur as singular or composite systems.

Combined Drain-cum-Sub Irrigation System (CDSI System)

Sometimes, combined surface and subsurface drainage systems are used. Whether this is needed or not depends on a combination of factors: the intensity and duration of the rainfall, surface. Storage, the infiltration rate, the hydraulic conductivity and the groundwater conditions. Some examples of combined systems are:

From the functional point of view surface drains can be further categorized into: (a) those that carry only the surface water from storms, (b) those that carry only the seepage water draining out of the soil mantle, and (c) those that combine drainage from both the sources.

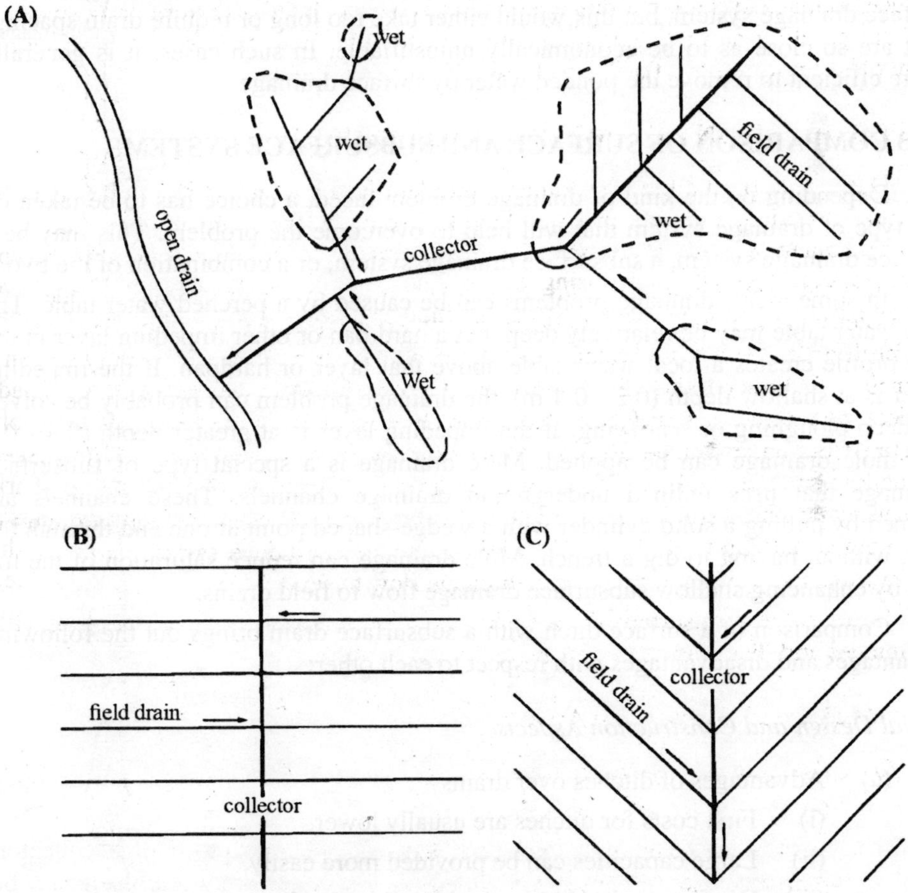

Figure 1.2. Combined Drainage System

Likewise, the subsurface drains can be grouped based on their alignment with respect to the direction of the movement of ground water that gets drained. Thus, *relief drains* are those that are located at suitable depth and spacing so as to attract water below the water table into them all along their length as spatially varying flow. *Intercepting drains* have their alignment perpendicular to the general direction of the flow of ground water. *Ecodrains* are plants and trees that are grown to effect control over water table elevation — this is a recent concept considered to be ecologically attractive.

In irrigated areas in arid and semi-arid regions, where the cropping pattern includes rice in rotation with dry-foot crops. Subsurface drainage is needed to control salinity for the dry-foot crops, whereas surface drainage is needed to evacuate the standing water from the rice fields (e.g. before harvest).

Areas with occasional high-intensity rainfall (say more than 50 mm/day), which causes water to pond at the soil surface, even when a subsurface drainage system has been installed.

In both of these examples, the standing water could be removed by the subsurface drainage system, but this would either take too long or require drain spacings that are so close as to be economically unjustifiable. In such cases, it is generally more efficient to remove the ponded water by surface drainage

1.13 COMPARISON OF SURFACE AND SUBSURFACE SYSTEMS

Depending on the kind of drainage problem faced, a choice has to be taken on the type of drainage system that will help to overcome the problem. This may be a surface drainage system, a subsurface drainage system, or a combination of the two.

In some areas, drainage problems can be caused by a perched water table. The true water table may be relatively deep, but a hard pan or other impeding layer in the soil profile creates a local water table above that layer or hardpan. If the impeding layer is at shallow depth (0.2 - 0.4 m), the drainage problem can probably be solved by deep ploughing or scarifying. If the impeding layer is at greater depth (0.4 - 0.8 m), mole drainage can be applied. Mole drainage is a special type of subsurface drainage that uses unlined underground drainage channels. These channels are formed by pulling a solid cylinder with a wedge-shaped point at one end through the soil, without having to dig a trench. Mole drainage can reduce saturation of the top soil by enhancing shallow subsurface drainage flow to field drains.

Comparison of a surface ditch with a subsurface drain brings out the following advantages and disadvantages with respect to each other:

Initial Design and Construction Aspects

(a) Advantages of ditches over drains -

  (i) First costs for ditches are usually lower,

  (ii) Large capacities can be provided more easily,

  (iii) Ditches can be designed with milder gradients.

(b) Disadvantages compared to subsurface drains are as follows-

  (i) Ditches occupy a lot of valuable land,

  (ii) Ditches have relatively limited effectiveness in drawing down the ground water level and also leaching of soil salinity,

(iii) Structures for crossing to be used for the movement of men and machinery have to be provided.

The only major advantage that surface ditches have over subsurface drains is that the ditches are open for inspection, and direct access is available for repairs. The major disadvantages of ditches are problems of weed growth and ploughing of sides and the resulting high cost of maintenance.

CHAPTER - 2

DRAINAGE INVESTIGATION

High watertable and salt problems often develop as a result of canal irrigation of level lands, which may be well drained under natural conditions. They may have drainage facilities inadequate to take care of the additional groundwater level may raise from over irrigation. In that event the groundwater level may rise from considerable depth within the root zone in a relatively short time. The sources of waterlogging are seepage from canals, hillsides or higher irrigated areas accumulation of monsoon rainfall in low-lying areas for want of suitable outlet, and over-irrigation. It may be surprising that about one centimeter of excess water which may percolate below the crop root zone and reach the groundwater may raise that water table about 8 to 10 centimeters, even in unirrigated areas drainage problems may arise in level valley lands due to seepage from surroundings hillsides. Seepage from canals in some areas may account for as much as 10 to 30% of the total water delivered to the canal. It may be pointed out that the effects of these losses are frequency not felt in the areas in which they occur but may manifest themselves many kilometers away from their source. For example, the seepage from canal or excess water from an irrigated area may seep down through the ground in lateral aquifers to reappear 5 or 10 kilometers from the area in which it is applied.

Drainage Investigation provides a good deal of information on the possibilities of a suitable outlet, area of the watershed contributing runoff the low lying area, rainfall characteristics, extent and severity of the drainage problem, cost of land, soil type and the type of crops which may grow in the area. A topographic map of the problem area and the surrounding watershed will provide information on the characteristics of runoff into the area, possible location of the drains and allied structures including culverts and structures, and preliminary estimates of the cost of the drainage system. The characteristics of groundwater flow and the extent of water logging in the problem area are determined from the measurement of the water level in observation wells, or piezometers established as observation well readings could give the direction of groundwater flow and depth to the watertable.

2.1 BASIC DATA REQUIRED

The basic data must be sufficiently representative to permit selection of good drainage plan from which a functionally sound drainage system can be designed and constructed. Inadequate or unreliable data introduce serious risks in determining the drainage requirements and cost estimates. The basic data must provide a knowledge of (i) capacity of the soils to transmit water, (ii) amount, source, movement and chemical characteristics of the water that must be transmitted; and (iii) available hydraulic gradients, both natural and those induced by man. Sufficient data must be gathered to estimate the effects of the drainage plan on both the social and economic environment.

2.2. COLLECTION OF EXISTING DATA

A number of different agencies collect field data which are relevant to drainage design. Therefore, first prepare a list of those agencies and the available data with them. Some of the Departments which can be contacted for this purpose are; Revenue, Agriculture, Irrigation and Drainage Departments of the States, Central and State Groundwater Boards, Survey of India, Minor Irrigation Corporations, Meteorological Observatories and other research organisations located in the area. The information available from these Departments could be of following type.

i. topographic maps - survey of India and revenue maps and areal photographs

ii. Soil survey and geological reports

iii. Well logs

iv. Water table fluctuations

v. Irrigation and drainage maps

vi. Climatological and meteorological data

vii. Information regarding irrigation methods

viii. Cropping pattern and crop yields

ix. Size of land holding and ownership.

2.3. TYPES OF INVESTIGATIONS

Investigations of a drainage project can be mainly divided into following three groups:

Reconnaissance survey

The first exploration should be intended to have a quick overall look at the area. For reconnaissance survey a map of not smaller than 1/25,000 or 1/50,000 scale is generally adequate. In any reconnaissance survey additional area around the project region should always be included. The survey consists of moving around the area, inquiries from local people and' to locate on the base map information about the physical boundaries, surface water boundaries, cropping conditions, source of water for irrigation and its location, depth to water table in existing wells in the area, any mark of mottling indicating the highest and lowest water table heights, and the position and availability of natural drainage and outlets. The existing grades, excessive erosion areas, natural side slopes of existing channels, flows in drains and existing sites of culverts and bridges are also noted at appropriate place. High water marks on old buildings, bridges and stream banks will generally reveal the severity of floods in the area.

The major aim of this survey is to get familiar with the area and simultaneously get as much information as can be gathered. Once the reconnaissance survey is completed, more detailed surveys are undertaken. The care with which the reconnaissance survey is made will have considerable influence on the other surveys and will determine the amount of time and money that can be spent on various items.

Preliminary survey

Preliminary survey is an initial step for collecting broad details of proposed project area so as to enable preparation of a feasibility project report. It should include the following:

- i. area shape and land use in the watershed
- ii. rainfall data
- iii. topographic and physical features of the command area
- iv. watertable depths and fluctuations
- v extent of waterlogged and salt affected areas (these should also be marked on copies of revenue maps)
- vi. a study of the present cropping pattern and yields as well as the level of soil productivity

Whole surveying the physical features of the command area, the following aspects should be looked into:

- i. outlet position
- ii. topography including natural channels, depressions and lake level etc.
- iii. size of the area
- iv. texture and physical condition of the soil
- v. bridges, highways, railway line, and canals
- vi. property boundaries
- vii. present land use.

These data should be in sufficient detail to provide a basis of making tentative designs, estimates and financial/economic feasibility of the projects.

The waterlogged and salt affected areas are found scattered all over the project command area, and are shown on survey map of the whole . project. In addition, these areas are plotted individually for each drainage zone on the copies of revenue maps, which provide local features and the names of villages. However, these maps might have been prepared many years ago and hence a chain compass survey should be carried out to plot the exact course of the natural steams, nallahs, and drains and to show accurately the waterlogged and salt affected areas, field numbers, irrigation channels, cart tracks, walls, bridges etc.

Design survey and investigations

The purpose of the design survey and investigations is to provide all relevant data required to prepare a detailed project report (DPR). Hence all activities are aimed at collecting detailed information for the project area from the following surveys and investigations.

Topographic survey: It is the basis for further investigations and therefore data collection should begin with preparing a detailed topographic map of the area. A large scale map on 1:2400 to 1:6000 scales should be appropriate. Whereas for land drainage, minor field elevation differences are not important, they are to be accounted for while designing a field drainage system contour interval in the range of 15 cm upto 1 in 1000 slope and of 30 cm in higher sloping topography should show all topographic and physiographic features relevant to drainage apart from elevations of the fields, elevations of all critical locations such as water elevation in existing drains, roads, railways, or pipe lines, farm boundaries, land use pattern etc. shall be noted. Such information will add to the value of the map.

2.4. SOIL SURVEY

Detailed information on soils of the area to be drained Is basic to any drainage design. A whole profile in situ to be depth of 2-3 meters needs to be examined. Standard soil survey is adequate for planning a surface drainage system. For sub surface drainage system usually greater details are needed. The following characteristics are important from drainage point of view :

- texture
- structure, particle size distribution and aggregation
- salinity and alkalinity
- depth and suitability for cropping

The soil texture, structure, permeability and infiltration characteristics of the soil etc. effect the process of drainage.

Soil having good permeability do not require drainage except when the soil has a high water table. Slowly permeable and level and receiving high rainfall often need drainage practices to be adopted.

Texture of the soil effect the pore size distribution and drainage properties of the soil. Fine textured soil like clay has smaller size of pores and thus need great effort for drainage. Similarly a coarse textured soil which has larger pore space needs less efforts for drainage. The need for drainage efforts is reduced as the particle size of the soil increased from clay to silt, loam, sandy load and sand.

Soil structure i.e. arrangement of soil particles effects considerably the drainage properties of the soil. Silt having platy structure, or aggregates with horizontal axis longer than the vertical have poor drainage properties. They resist downward movement of water. Prismatic structure aggregates with their horizontal axis shorter than the vertical, permits easy downward movement of water. Blocky structure having aggregates with horizontal and vertical axis more or less equal and also the granular structure where the aggregates are nearly round offer favorable drainage characteristics.

Infiltration characteristics of the soil surface and permeability of the lower horizons greatly effect the drainage properties of soil.

Information of Soils for drainage

Soil is the medium in which plant roots anchor and from which they extract water and nutrients; soil is a matrix with particular characteristics of water entry and permeability, soil is that part of the earth's crust where soil has formed as a result of various interactive processes. This section discusses the pedological base of soil formation. The consistency of soil refers to the effect of the physical forces of cohesion and adhesion within the soil at various water contents. The terminology used ranges from loose to extremely hard in dry soil, from loose to extremely firm in moist soil, and from non-sticky, non-plastic to Very sticky, very plastic in wet soil. The arrangement of soil particles-affects the downward movement of water. A platy structure, aggregates with horizontal axis longer than the vertical, have poor drainage properties as they resist the downward movement of water. Prismatic structures having aggregates with their horizontal axes shorter than the vertical, permits easy downward movement of water. Blocky structure having aggregates with horizontal and vertical axes more or less equal and also granules structure where the aggregates are nearly round, offer favorable drainage characteristics. Infiltration characteristics of the soil surface and the permeability of the lower horizons greatly affect the drainage properties of soils. Much can be learned about a soils internal drainage by digging into the soil. Permeable soils that do not require drainage are uniform in colour through the profile. Yellow sub-soils usually indicate intermediate permeability, especially in day soils. Sub-soils, which are mottled with red, yellow and Grey are more slowly permeable than yellow sub-soils. Grey clay sub-soils in humid regions indicate very slow permeability.

A soil may need artificial drainage to lower down the high water and to remove excess surface water that cannot move downward through the soil to prevent the plant roots from suffocating. A permeable soil seldom needs artificial drainage, except when the soil has a high water table. Slowly permeable soils often need artificial drainage, especially when the land surface is level and rainfall is high. Texture or particle size affects the pore size distribution and the drainage property of soil. It is the predominance of small pore space, which causes the restriction to drainage. The finer particles having the smaller pores and greater the need for artificial drainage. Conversely, the need for removal of excess water by artificial drainage is reduced as particle size increases from clay to silt to loam to sandy loam to sand. Other aspects of soil, such as water retention, workability, and fertility, strongly affect plant productivity, and need to be assessed or studied in conjunction with drainage.

Hydraulic Conductivity of soil

The term hydraulic conductivity is the proportionality factor K in Darcy's law for the flow of water in soil.

$$v = -Ki$$

V - Flow velocity, LT^{-1}

K = Hydraulic conductivity, LT^{-1}

i = hydraulic potential gradient (dimensionless)

The negative sign indicates that the flow is in the direction of decreasing head. Darcy in 1856 observed that the amount of water flowing through the sand sample in unit time (as in other words the rate of flow as the discharge) was proportional to the difference Ah between the fluid heads at the inlet and outlet faces of the sample (head loss $h_1 - h_2 = \Delta h$), and inversely proportional to the length of the sand sample (the flow path). This proportionality can be expressed mathematically as follows:

$$Q = K\frac{\Delta h}{L}A$$

Where,

Q = Rate of flow through the sample

Δh = Head loss m (L)

L = Length of sample m (L)

A = Cross sectional area of tube m^2 (L^2)

K = Proportionality constant, depending upon the nature of sand and the fluid (LT^{-1})

The proportionality factor, K, is known as the coefficient of permeability or hydraulic conductivity. It can be seen that for unit hydraulic gradient the flow velocity equals K. The factor K usually stands for the hydraulic conductivity of saturated soil, i.e. under positive pressure of the soil water. It is considered independent of the pressure gradient because we assume that the water-conducting porosity does not charge with pressure gradient. Under unsaturated conditions, however, the hydraulic conductivity varies with soil moisture suction, since suction, soil moisture content, and water-conducting porosity are closely interrelated. The hydraulic conductivity of a soil represents its average water transmitting properties, which depends mainly on the number and the diameter of the pores present

Homogeneous soil- If the number and diameter of the pores are distributed uniformly, the soil is said to be homogeneous. *Isotropic soil:* If the hydraulic concavity is same in all direction the soil is said to be isotropic. *Anisotropic soil-* Soil commonly shows certain stratification so that K in one direction is greater than in another. A soil in which hydraulic concavity at any point has perfect direction is called Anisotropic soil. *Heterogeneous Layer:* If the anisotropy varies from point to point in a given layer the layer is said to be Heterogeneous Layer.

Hydraulic conductivity measurements in field can be expected to yield more reliable results than those on disturbed soil samples. Table gives qualitative estimates of hydraulic conductivity as influenced by soil texture.

Table 2.1. Soil Texture - Hydraulic Conductivity Relationship

S. No.	Soil texture	Hydraulic conductivity, m/day
1	Clay	<0.01
2	Silty clay	0.010-0.015
3	Silty clay loam	0.015-0.024
4	Clay loam	0.024-0.012
5	Sandy clay loam	0.012-0.091

S. No.	Soil texture	Hydraulic conductivity, m/day
6	Silt loam	0.091-0.213
7	Loam	0.213-0.427
8	Very fine sandy loam	0.427-0.732
9	Fine sandy loam	0.732-1.036
10	Sandy loam	1.036-1.524
11	Loamy fine sand	1.524-2.134
12	Loamy sand	2.134-2.743
13	Fine sand	2.743-3.658
14	Sand	3.658-4.877
15	Coarse sand	4.877-7.315

Soil-water Relationship

In a soil, the solid phase usually controls the form or spatial distribution of the liquid and the gas phases. The solid phase is therefore called the soil matrix.

Over most of the wetness range in which plant roots normally function, properties of soil-water retention and transmission area determined by forces associated with the soil matrix. Interactions between the soil matrix and the water are basically due to the forces of adhesion and cohesion.

The availability of soil water related to its energy status, which is referred to as the water potential. The water potential is governed by the matric forces and by the force of gravity. Other factors may also affect the water potential: the osmotic pressure of dissolved salts, the external gas pressure, and the pressure arising from the swelling of clay. For our purposes, we define the water potential as the sum of the matric potential and gravitational potential.

The existence of the matric potential can be demonstrated by means of a tensiometer placed in the soil. Provided the soil is not saturated, water will move from the porous cup of the tensiometer into the soil. At equilibrium, a negative pressure is measured on the tensiometer.

The elevation head depends on the difference between the level of the point where we define the energy status of the water, and a reference level. Usually, the watertable is taken as the reference level.

Above the watertable, the matric head has a negative value because work is needed to extract soil pores against the action of the matric forces. This requires a negative pressure or suction. The matric forces decrease sharply when the radius of the pres increases. The matric head is a function of the soil-water content.

Imagine that free drainage occurs in a soil that has become saturated after a heavy rainstorm. If the soil has large pores in which the matric forces are small, these pores will release water by gravity flow. After this water is released, the soil is at field capacity, corresponding with a volumetric soil-water content at a matric head somewhere between -100 and -200 cm ($2.0 < pF < 2.3$). The soil-water content will further decrease by cop transpiration and evaporation at the soil surface. The remaining soil water redistributes by flow through capillaries and flow along the walls of empty pores. When the matric head $h_m = -16000$ cm (pF = 4.2), the soil is at wilting point, because plant roots cannot extract water from the soil when the matric head falls below this point The soil water stored between field capacity and wilting point is called the 'available soil water' or the soil's water holding capacity. Usually, pF-curves are measured by the stepwise drying of a wet sample (desorption). When the watertable is at shallow depth, the matric head at field capacity is less well defined, because, if the watertable influences the soil-water conditions in the root-zone, free drainage will be prevented. If the watertable is lowered, a certain amount of water in the unsaturated part of the soil profile will be released by gravity flow.

2.5. WATER SOURCES SURVEY

Any remedy through drainage can only be accomplished satisfactorily if it is possible to pin point the source of the excess water in the area. The excess water can be due to one or a combination of several factors. Accumulation of rain water in flat areas requires a study on frequency analysis of rainfall, adequacy of natural drains and outlets, obstructions in such courses and top soil conditions and storage capacity of the soil profile.

Periodic flooding of the land due to overflow of rivers, branches of canals of drainage streams, the location and extent of flooding, the frequency, depth and duration of flooding, time of the year and harmful side effects of the flooding should be investigated.

Problems arising out of runoff from areas outside, the project calls for a study on the frequency analysis of rainfall characteristics as well as of the watershed as a whole. This should enable the planner to decide whether up stream regulatory works are needed or a diversion drain is needed to relieve the problem. However, flood control measures are not considered part of the agricultural drainage of irrigated areas.

The problems of waterlogging in low lying areas could also arise due to seepage from irrigated upland. Interceptor drains could be thought of as an ameliorative step.

High water table due to seepage from irrigation water distribution network and deep percolation losses needs investigation on areas of ground water recharge, extent of recharge, transmission and discharge. Natural drainage component should be thoroughly investigated as it can be reduce drainage costs considerably by reducing drainage coefficient.

Development of waterlogging due to artesian pressure requires investigations on the feasibility of tube well drainage as narrow spacing for pipe drainage in such cases will make the proposal, uneconomical.

2.6. GROUND WATER INVESTIGATIONS

Depth to watertable and its seasonal and annual fluctuations determine the extent and severity of drainage problem. For this purpose, observation wells are installed mostly on a square grid pattern. The density of observation net work depend upon the topography as well as geological and hydrological conditions of the area. Table 1 gives rough guide to estimate the number of observation wells.

It is generally useful to include additional areas beyond the actual project area (up to the watershed line) in any ground water investigation.

Groundwater Data Collection

Depth to watertable and its seasonal and annual fluctuations determine the extent and severity of drainage problem. For this purpose, observation wells are installed mostly on a square grid pattern. The density of observation net work depends upon the topography as well as geological and hydrological conditions of the area. For the collection of data on the depth and configuration of the watertable, the direction of groundwater movement and the location of recharge areas, a network of observation wells and/or piezometers has to be established.

Data collection from the existing wells

In many villages and farms have shallow, hand-dug wells that can offer excellent ground water observation points. Because they are hand-dug, one can be sure that they will not penetrate below the lowest level to which the groundwater will fall. They will thus truly represent the watertable. Their location, however, may not fit into an appropriate network; they may be sited at topographic highs, for example, or their water levels may be deeper than 2m below the land surface. Due to large diameter they have a large storage capacity, implying that the water level in the well will take some rime to respond to changes in the watertable, or to recover when water is taken from them in substantial quantities. Before existing wells are included in network of observation wells, therefore, information should be collected on their depth, diameter, construction, layers penetrated, and frequency of use.

2.7 PIEZOMETERS AND OBSERVATION WELLS

Piezometers

Piezometer means "pressure meter" is a small diameter, usually 2.5 cm pipe driven vertically into the ground surface. A piezometer is a small diameter pipe, driven into, or placed in, the subsoil so that there is no leakage around the pipe and all water enters the pipe through its open bottom. Piezometers are installed to indicate the hydrostatic pressure of ground water at the lower end of the pipe. Water enters the piezometers through the open bottom. There is no leakage through the sides. When the soil strata contains permeable and impermeable formations near the ground surface, reliable information on the characteristics of the ground water can be obtained by installing piezometers. At the end of the piezometer, there should be a close fit of the pipe in the soil. Piezometers may be installed by driving or by jetting.

A battery of piezometers is first provided at one location in which the lower end of each piezometer terminate at different depth. Their number is increased if the problem of artesian pressure is indicated. Measurements are undertaken periodically once or twice a month. From the difference between any two readings spread over a period of one or more years, trend of water table rise or fall can be estimated.

a. The piezometer indicate the ground water is going down and there is some natural drainage

b. The piezometer indicate a hydrostatic pressure or that there is water coming up from a deeper strata

c. The piezometer indicate that hydrostatic pressure in a stratum, and that water is being forced up and down the stratum

d. The piezometers indicate that ground water is moving into a stratum.

Installation of Piezometers

To install piezometers by driving, the pipe is fitted with a driving head to protect the upper end from the hammer blows and the lower end is plugged with a round-headed rivet. The rivet can be forced deeper in the soil by driving with a steel rod when the piezometer is at the required depth, leaving the end of the pipe open.

Piezometers are driven vertically into the ground. A circular centering table having a metal frame and a 60 cm diameter wooden top provide a convenient device to drive the pipe vertically into the ground. The table is placed level on the ground with the help of level bubble tube. The pipe is driven by hammer. After the pipe is driven to the full depth, soil inside the pipe is the taken out by a screw auger. The end is not provided with a plug (rivet) while driving.

For the successful operation of piezometers, the soils in the lower end of the piezometer are removed by flushing to form a small cavity of about 15 cm depth at the bottom of the pipe.

The equipment required for flushing consists of a portable piston pump, a water drum and enough plastic tubing to reach the bottom of the deepest piezometers.

A point is marked on the tubing which indicates the distance the tubing has to be inserted while developing the cavity. This distance is the length of the pipe plus the depth of the cavity.

The piezometers are flushed once every six months or more frequently if required to ensure that they are not plugged and rendered inoperative.

Piezometers installed by jetting require a high pressure, low volume pump, and a supply of water. A high pressure stream of water is pumped through a hose connected to the piezometer pipe. The water prevents the pipe from being plugged with soil and at the same time washes an opening ahead of the pipe into which it can be pushed.

Piezometers and observation wells are usually installed on a grid pattern extending the entire farm and the area adjoining the farm contributing seepage into it.

The grid spacing varies between 3 and 300 m or more, depending on the precision required in the survey. At each grid point two or more piezometers are installed at varying depth, depending on the stratification of the formation. The minimum space between individual piezometers in a battery is 60 cm.

Observation Well

An observation well is a hole dug or augered into the ground for the purpose of observing the fluctuations in water table. In an observation well, water enters through the entire saturated zone and indicates the water-table.

Construction of observation well:

- A 3 m length of 50 mm or 90 mm PVC pipe is ideal.
- A series of slots or holes are made in one end of the pipe for approximately 1 meter.
- A hack saw or 2.5 mm drill can be used to make the slots or holes.
- A commonly used size of the perforation is 3 mm diameter. The perforations are spaced to provide about 20% open area of the screened portion of the observation well.
- Residue must be cleaned out after the slots are made.
- An end cap should be placed on the end where the slots were made.
- A 2.5 mm hole should be drilled in the bottom of the cap to allow for drainage.

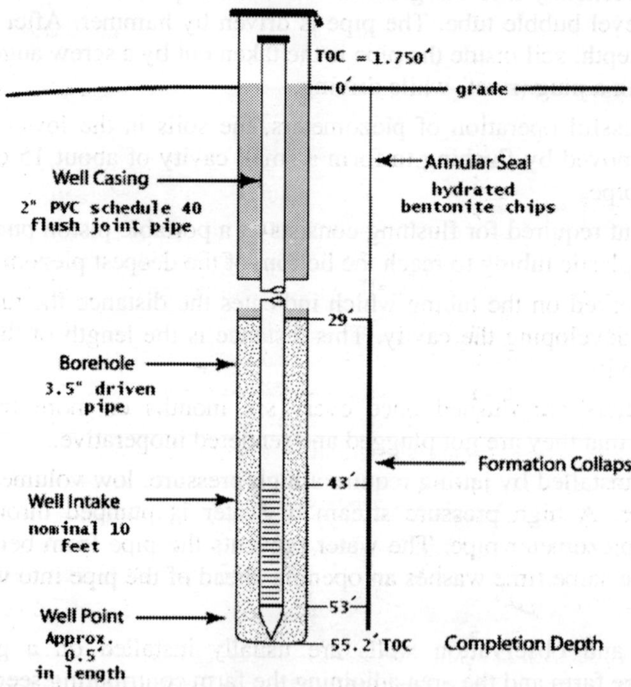

Figure 2.1. Observation Well

Installation of Observation wells

- Use a post-hole digger or 100 mm auger to dig a hole 300 mm less than the length of the pipe. A 2.5 cm diameter galvanized iron pipe is usually used in making observation wells. Ensure that the hole reaches, or has penetrated any relatively impermeable rock or clay material.
- Place the pipe in the hole and pack clean, well-graded gravel up the outside of the pipe to a level well above the slots. Place clay or fine textured soil on

top of the gravel to fill the hole almost to ground level. This should be compacted or tamped using a length of steel rod or similar.

- Gravel packing is usually provided for observation wells located in formations having fine sand. To provide for gravel packing, in an observation well of 2.5 cm diameter, a 5 cm diameter auger hole is first drilled and the well pipe installed in it.
- The annular space is filled with gravel of suitable size to retain about 90% of the formation material.
- When the formation material comprises predominantly of coarse particles a coir strainer with coir rope wound around the perforated portion of the observations well pipe may be used with advantage.
- Finally, pour cement slurry or compact clay around the pipe at ground level. This will stabilize the observation well and prevent water flowing down the outside of the pipe.
- A removable cap or tin can may be placed on top of the pipe to prevent vermin entering and clogging the pipe.

2.8 OBSERVATIONS

Measurements are made with an electrical depth gauge or a tape with a chalked end. Observations are made periodically on the farms. The elevation of the ground surface (RL) at the piezometric and/or observation well stations is established with reference to a permanent bench mark on the farm. The height of the pipe above the ground surface is noted. Observation wells and piezometers are read by measuring the distance from the top of the pipe to the water surface and then subtracting the length of the pipe above the soil surface. Attach a weight (sinker or galvanized plug end) to the end of a tape measure. Lower the weight into the well and listen for the splash as the weight hits water level. Read and record the distance from water level to the top of the well casing. Subtract 300 mm (or the length of pipe protruding above ground level) from your reading to give you the distance between ground level and the water level. Record all measurements including the date and well location. Also note any rainfall events and irrigation details.

Uncased wells and cased wells

The uncased wells can be used successfully in soils that are enough to prevent the borehole from collapsing. They can easily be made with a hand auger as used in soil surveys, and can be 50 to 80 mm in diameter. They are also a cheap means of measuring watertable levels, when the primary objective is to obtain a rough idea of the groundwater conditions in the project area. When making an observation well in unstable soil, one has to use a temporary casing, say 80 or 100 mm in diameter. The casing prevents sloughing and caving and makes it possible to bore a hole that is deep enough to ensure that it always holds water. Whatever casing material is locally available can be used: sheet metal, drain pipe, or standard commercial types of well casing

The borehole is made by hand auger or other lightweight boring equipment until the watertable reached. After lowering a casing, at least 80mm in diameter, into the hole, one deepens the hole by bailing the material inside the casing. When there is difficulty in keeping the sand from heaving inside the casing, this can be overcome by adding water to keep the water level in the pipe above the water level in the water-bearing layer. When the borehole has reached the required depth, a pipe at least 25mm in diameter is then lowered inside the casing to the bottom of the hole. Centering this pipe in the casing is important. The pipe's lower end must contain slots and perforations over a length equal to the distance over which the watertable is suspected to fluctuate. The next step is to fill the space between the pipe and the casing with graded coarse sand or fine gravel up to some distance above the upper limit of slots or perforations; the remaining annular space can be backfilled with parent materials. A properly placed gravel pack facilitates the flow of groundwater into the pipe, and vice versa, and prevents the slots or perforations from becoming clogged by fine particles like clay and silt. Finally, one pulls out the casing and places a concrete slab around the pipe to protect it from damage. Wrapping a piece of jute or cotton around the perforated part of the pipe may then suffice. It is advisable to remove any muddy water from the completed well by bailing.

Layout for observation

To save costs, observation wells and piezometers should be installed concurrently with soil borings that arc needed to explore the shallow subsurface. These borings are usually made on a rectangular grid pattern that is laid out on the basis of information on topography, geology, soils, and hydrology collected during the early phase of the project. Soil borings should be spaced rather close together to make it possible to correlate subsurface layers. It is not necessary to transform each soil boring into an observation well because the watertable is a smooth surface. A network of observation wells should be: along and perpendicular to, lines of suspected groundwater flow; at locations where changes in the slope of the water-table occur or are suspected; on the banks of streams or other open water courses and along lines perpendicular to them; on areas where shallow water tables occur or can be expected in the future (areas with artesian pressure and areas with a high intensity of irrigation); along and perpendicular to the (project) area's boundaries.

Density and depth of wells

The density of wells depends on the topographic, geological, and hydrological conditions of the area and on the type of survey. As the required accuracy is generally inversely proportional to the size of the area. The depth of observation wells should be based on the expected lowest groundwater level. This will ensure that the wells do not fall dry in the dry season and that readings can be taken throughout a full hydrological year. The lowest water level can only be estimated, unless data from previous investigations are available. Generally, water tables deeper than about 3 m are not interesting from the viewpoint of planning a drainage system. Observation wells to this depth are therefore adequate in most flat lands. In areas with a rolling topography, deeper observation wells may be needed on the topographic highs to obtain a complete picture of the groundwater conditions. In stratified soils and particularly in areas where artesian pressure exists or can be suspected, a number of deep piezometers are needed in addition to the shallow ones.

In many alluvial plains, the covering layer is made of alternating layers of heavy and light-textured materials, or even peat. The total thickness of this layer can be many metres. In this case, a multi-piezometer well can best be made, containing 3 to 5 piezometers placed at different depths (e.g. at 2,5, 8,12, and 15 m). The deepest piezometer should be placed in the underlying coarse sand layer, where groundwater is under artesian pressure.

Well Elevations

To determine the elevation of the observation wells and piezometers and thus be able to correlate watertable levels with land surface levels, a leveling survey must be made. Water levels in the wells are measured from a fixed measuring point, which, for cased wells, can be the rim of the casing; for uncased wells, a measuring point must be made (e.g. a piece of steel, wood, or stone).

2.9. METHODS FOR MEASURING WATER LEVELS

Wetted tape method

A steel tape (calibrated in mm), with a weight attached to it, is lowered into the pipe or borehole to below the water level. The lowered length of tape from the reference point is noted. The tape is then pulled up and the length of its wetted part is measured (This is facilitated if the lower part of the tape is chalked). When the wetted length is subtracted from the total lowered length, this gives the depth to the water level below the reference point.

Mechanical sounder

This consists of a small steel or copper tube (10 to 20 mm in diameter and 50 to 70 mm long), which is closed at its upper end and connected to a calibrated steel tape. When lowered into the pipe, it produces a characteristic plopping sound upon hitting the water. The depth to the water level can be read directly from the steel tape.

Electric water-level indicator

This consists of a double electric wire with electrodes at their lower ends. The upper ends of the wire are connected to a battery and an indicator device (lamp, mA meter, and sounder). When the wire is lowered into the pipe and the electrodes touch the water, the electrical circuit closes, which is shown by the indicator. If the wire is attached to a calibrated steel tape, the depth to the water level can be read directly.

Floating level indicator or recorder

This consists of a float (60 to 150 mm in diameter) and a counterweight attached to an indicator or recorder. Recorders can generally be set for different lengths of observation period. They require relatively large pipes. The water levels are either drawn on a rotating drum or punched in a paper tape.

Pressure logger or electronic water-level logger

This measures and records the water pressure at one-hour intervals over a year. The pressure recordings are controlled by a microcomputer and stored in an internal, removable memory block. At the end of the observation period or when the memory block has reached capacity, it is removed and replaced.

2.10 FREQUENCY OF OBSERVATIONS

The watertable reacts to the various recharge and discharge components that characterize a groundwater system and is therefore constantly changing. Important in any drainage investigation are the (mean) highest and the (mean) lowest watertable positions, as well as the mean watertable of a hydrological year. For this reason, water- level measurements should be made at frequent intervals for at least a year. The interval between readings should not exceed one month, but a fortnight may be better. All measurements should, as far as possible, be made on the same day because this gives a complete picture of the watertable. Recorded for each observation are: date of observation, observed depth of the water level below the reference point, calculated depth below ground surface (for free water tables only), and calculated water-level elevation (with respect to a general datum plane, e.g. mean sea level). Other particulars should also be noted (e.g. number of the well, its location, depth, surface elevation, reference point elevation). If one wants to study the effect that a rain shower or an irrigation application has on the watertable, daily or even continuous readings may be needed. A pressure logger or an automatic recorder should then be installed in a representative large-diameter well; depending upon the type of recorder selected the well should have a certain minimum diameter, e.g. 7cm.

2.11. ANALYSIS OF DATA

Hydraulic Gradient

The hydraulic gradient can be measured by installing several piezometers side by side, but at different depths below the soil surface. If two pipes of unequal length are placed vertically in a static body of water so that the tops of both are at the same distance above the water surface, the water level inside the pipe will be at the same distance below the top of both the pipes. The deeper pipe ends a zone of higher pressure because of its depth, and this pressure is measured by the height of the water column from the bottom of the pipe. The shallower pipe ends in a zone of lesser pressure and, as a result, has a shorter water column. As the body of water is static, the water level, as measured from the top of both the pipes, is the same, which indicates no movement.

The situation for piezometers installed in the soil is the same. If the water level, as measured from the top of two or more pipes of unequal length, which project the same distance above the ground surface, is the same, there is no vertical component of flow in the soil. When any two adjacent piezometers are compared and the water level in the deeper one is higher than that in the shallower one, an upward gradient is indicated. Conversely, if the water level in the deeper pipe is lower than that in the shallower pipe, a downward gradient is indicated. There may be horizontal groundwater movement at a piezometer battery, which shows no vertical component of flow or vertical gradient. This can be determined by installing another battery of piezometers at a different location. If the water level at one battery is higher than at the other, flow from the higher toward the lower is indicated.

Depth to impermeable layer

Whether the soil profile is homogenous or layered, the presence or absence of a pervious horizon at or below drain depth, presence or absence of impeding horizons and the depth to impervious base are generally investigated during soil survey. Depth to impervious base is, however, discussed here as it forms an important design parameter. As the location of drains with respect to ground surface and impermeable layer effect drainage design, most of the design formulae require information on this parameter.

An impermeable layer is not completely impervious to water as may be understood from the term it is a relative term. A layer is designated as impermeable if its hydraulic conductivity is less than 20 per cent of the weighted mean hydraulic conductivity of the upper soil layers. It is useful to remember that presence of an impermeable layer may be investigated upto 6 m using hand auger. In the absence of any impermeable layer upto this depth, well logs and other available data should be examined at least upto 13 m depth. Under most practical situations when drains are laid upto 2.5 m depths, any impermeable layer below this depth will have very little influence on drainage design. Sufficient data should allow plotting of the depth of the impermeable barrier. A contour map of pervious base layer should generally reveal the desired information. Occurrence of hard pan within 50-70 cm of the soil surface will probably prohibit use of pipe drainage.

Moisture characteristic curves

Moisture characteristic curves represent the volume of water retained by the soil at a particular suction. The drainable pore space can be evaluated by these curves from the known moisture content at saturation to that at 100 m suction. These curves are useful when soil moisture above the water table contributes to the flow when the water table is lowered.

Groundwater Hydrograph

A well hydrograph is prepared by plotting well water levels against time on a graph paper. If the water table is influenced by irrigation, the hydrograph shows a peak at the end of the irrigation season with subsequent falling trend. In monsoon climate regions, the peak occurs at the end of the monsoon season (Generally September/October) and minimum before the onset of monsoon (May/June). The response of water table rises or falls with the increase or decrease in the groundwater storage can be plotted in a hydrograph. Groundwater hydrographs show the water level readings, converted to water level below ground surface, against their corresponding time. A hydrograph should be plotted to each observation well or piezometer. In land drainage, it important to know the rate of rise of the water tabic, and even more important, that of its fall. If the groundwater is not being recharged, the fall of water table will depend on: the transmissivity of the water transmitting layer, KH, The storativity of this layer, S and The hydraulic gradient, dh/dx

After a period of rain (or irrigation) and an initial rise in groundwater levels, they then decline, rapidly at first, and then move slowly as time passes because both the hydraulic gradient and transmissivity decrease. The graphical representation the water table decline is known as the natural recession curve. It can be shown that the logarithm of the water table height decreases linearly with time. Hence a plot of the water table height against time on semi-logarithmic paper give a straight line. Groundwater recession curves are useful in studying the changes in groundwater storage and in predicting future ground water levels.

Chemical Properties of Soil and Water

The watertable may provoke the intrusion of salty groundwater from adjacent areas, or from the deep underground, or from the sea. The drained area and its surface water system will then be charged daily with considerable amounts of dissolved salts; The disposal of the salty drainage water into fresh-water streams may create environmental and other problems, especially if the water is used for irrigation and/ or drinking; In arid and semi-arid regions, soil salinization is directly related to the depth of the groundwater and to its salinity; Groundwater quality dictates the type of cement to be used for hydraulic structures, especially when the groundwater is rich in sulphates; Agricultural crops are affected by groundwater qualify if the groundwater approaches the root zone.

The weathering of primary minerals is the direct source of salts in groundwater. Bicarbonate is usually the primary anion in groundwater and forms as a result of the solution of carbon dioxide in water. Carbon dioxide is a particularly active weathering agent for such source rocks as limestone and dolomite. Sodium in the water originates from the weathering of feldspars, clay minerals, and the solution of evaporites. Evaporites are also the major natural source of chloride in groundwater, while sulphate originates from the oxidation of sulphide ores or the solution of gypsum and anhydrite. Such primary minerals as amphiboles (hornblende), apatite, fluorite, and mica are the sources of fluoride in groundwater. The mineral tourmaline is the source of boron. In the groundwater of coastal and delta plains, the sea is the source of salinity. Groundwater quality is also related to the relief of the area. Fresh groundwater usually occurs in topographic highs, which, if composed of permeable materials, are areas of recharge. On its way to topographic lows (areas of discharge), the groundwater becomes mineralized through the solution of minerals and ion exchange. Groundwater salinity varies with the texture of sediments, the solubility of minerals, and the contact time. Groundwater salinity tends to be highest where the movement of the ground-water is least, so salinity usually increases with depth. Irrigation also acts as a source of salts in groundwater. It not only adds salts to the soil, but also dissolves salts in the root zone. Water that has passed through the root zone of irrigated land usually contains salt concentrations several times higher than that of the originally applied irrigation water. Evapotranspiration tends to concentrate the salinity of groundwater. Highly saline groundwater can therefore, be found in arid regions with poor natural drainage and consequently a shallow watertable.

The choice of a method for measuring groundwater salinity depends on the reason for making the measurements, the size of the area- and thus the number of samples to be taken and measured -and the time and the budget available for doing the work-Once the network of observation wells, boreholes, and piezometers has been established, water samples should be taken in a representative number of them. Sampling can often best be combined with other drainage investigations, such as measuring hydraulic conductivity in open boreholes. The sample is then taken after a sufficient quantity of water has been bailed from the hole.

Electrical Conductivity

A rapid determination of the salinity of groundwater can be made by measuring its electrical conductivity, EC. Conductivity is preferred rather than its reciprocal, resistance, because the EC increases with the salt content. Electrical conductivity defines the conductance of a cm^1 of water at a standard temperature of 25°C. The salinity of groundwater can be rapidly determined by measuring its electrical conductivity (EC).

Electrical conductivity is expressed in deciSiemens per metre (dS/m), formerly in millimhos per centimetre (mmhos/cm). Expressing the results in terms of specific electrical conductivity makes the determination independent of the size of the water sample. Conductivity cannot simply be related to the total dissolved solids because groundwater contains a variety of ionic and undissociated species. An approximate relationship for most groundwater with an EC-value in the range of 0.1 to 5 dS/m is: 1 dS/m 640 mg/l.

The EC expresses the total concentration of soluble salts in the groundwater, but gives no information on the types of salts. These may be calcium, magnesium, sodium, potassium, carbonate, bicarbonate, chloride, sulphate, and nitrate, and need to be determined in the laboratory. Since these chemical analyses are costly, not all the observation points need be sampled for detailed analysis. A selection of sites should be made, based on the results of the EC-measurements.

pH

When the two electrodes are dipped into aqueous solution under test, the potential is develops in the solution. The difference in potential between both the electrodes is measured by pH meter.

2.12 GROUNDWATER MAPS

Watertable-Contour Map

Ground water contour maps or absolute depth of the ground water map are drawn by plotting elevations of the water table at a particular time on the field map. When superimposed on to a topographic map, it shows the relationship between ground surface and water table. Arrows are drawn on to the map to indicate the direction of ground water flow. It is expected that there is horizontal movement of groundwater in the direction of the slope of the water table. Radical changes in the direction of the slope may indicate unusual barrier formation perched water table, artesian conditions or high and low permeability areas leading to clues for further investigations. A water table contour map provides direct visual observation on the slope of the water table. To intercept ground water, drain lines are laid perpendicular to the direction of flow or parallel to the ground water contours. A water table-contour map shows the elevation and configuration of the water table on a certain date. To construct it, we first have to convert the water-level data from the form of depth below surface to the form of water table elevation (= water level height above a datum plane, e.g. mean sea level). A water table contour line is the locus of points on the water table for which the hydraulic head is constant. These data are then plotted on a topographic base map and lines of equal water table elevation are drawn. A proper contour interval should be chosen, depending on the slope of the watertable. For a flat water table, 0.25 to 0.50 m may suit; in steep water table areas, intervals of 1 to 5 m or even more may be needed to avoid overcrowding the map with contour lines. The topographic base map should contain contour lines of the land surface and should show all natural drainage channels and open water bodies. For the given date, the water levels of these surface waters should also be plotted on the map. To draw the water table-contour lines, we have to interpolate the water levels between the observation points, using the linear interpolation method as shown in Figure.

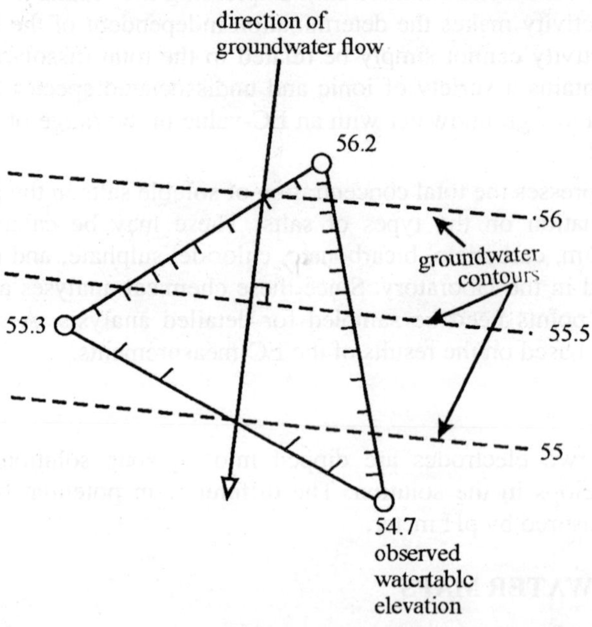

Figure 2.2. Construction of Water Table Contour Lines by Linear Interpolation

Instead of preparing the map for a certain date, we could also select a period (e.g. a season or a whole year) and calculate the mean water table elevation of each well for that period. This has the advantage of smoothing out local or occasional anomalies in water levels. A water table-contour map is an important tool in groundwater investigations because, from it, one can derive the gradient of the water table (dh/dx) and the direction of groundwater flow, which is perpendicular to the water table-contour lines. Water table contour map provides direct visual information on the slope of the water table. For artesian or irrigation areas in which two or more piezometers have been installed at the same location, with the bottom of each at a different depth in a different soil layer, contour maps of the hydraulic head in each layer should be made.

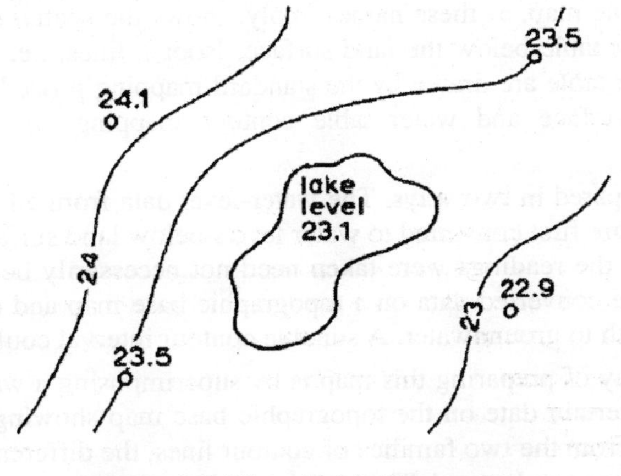

Figure 2.3. Water Table Contour Lines

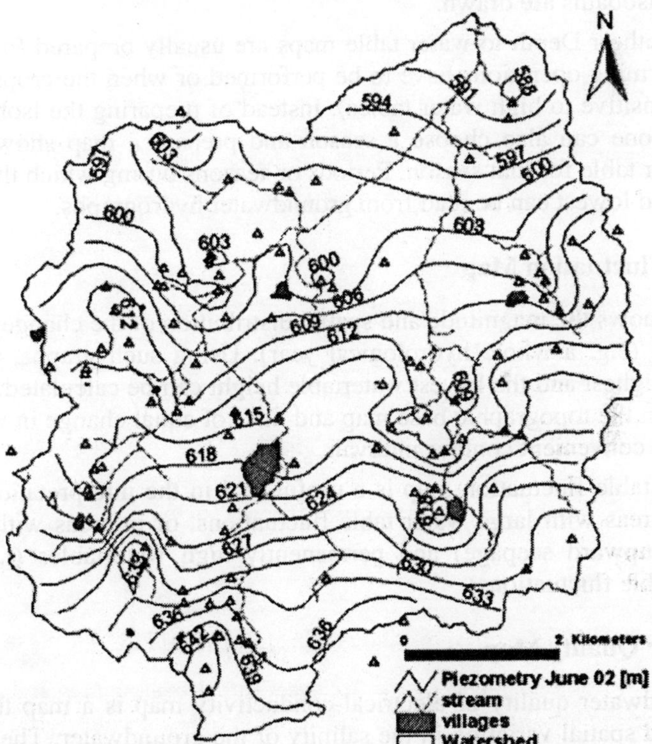

Figure 2.4. Water Table Contour Map

Isobath map or Depth to Water table Map

Depth to groundwater table below the soil surface are plotted on to a map and isocurves are drawn as on a contour map. An isobath maps is used to delineate areas where water table is critical (<1.5 m), marginal (1.5-3.0 m) and likely problem areas

(>3.0 m). There maps are generally used to justify a drainage project. An Isobath or depth-to-water table map, as these names imply, shows the spatial distribution of the depth of the water table below the land surface. Isobath lines, i.e. the lines of equal depth to the water table are drawn by the standard mapping procedures that are used for the ground surface and water table contour mapping i.e. interpolation and extrapolation.

It can be prepared in two ways. The water-level data from all observation wells for a certain date are first converted to water levels below land surface. The reference point from which the readings were taken need not necessarily be the land surface. One then plots the converted data on a topographic base map and draws isobaths or lines of equal depth to groundwater. A suitable contour interval could be 0.50 m.

The other way of preparing this map is by superimposing a water table-contour map made for a certain date on the topographic base map showing contour lines of the land surface. From the two families of contour lines, the difference in elevation at contour intersections can be read. These data are then plotted on a clean topographic map, and the isobaths are drawn.

An Isobath or Depth-to-water table maps are usually prepared for critical dates (e.g. when farming operations have to be performed or when the crops are expected to be most sensitive to high water tables). Instead of preparing the isobath map for a special date, one can also choose a season and prepare a map showing the mean depth-to-water table for that season. Periods or seasons during which the water tables are highest and lowest can be read from groundwater hydrographs.

Watertable Fluctuation Map

A map shows the magnitude and spatial distribution of the change in water table over a period (e.g. a whole hydrological year). Using such graphs, the difference between the highest and the lowest watertable height can be calculated. These data is then plotted on the topographic base map and lines of equal change in water table are drawn using a convenient contour interval.

A water table fluctuation map is a useful tool in the interpretation of drainage problems in areas with large water table fluctuations, or in areas with poor natural drainage (or upward seepage) and permanently high watertables (i.e. areas with minor watertable fluctuations).

Groundwater Quality Map

A groundwater quality or electrical-conductivity map is a map that shows the magnitude and spatial variation in the salinity of the groundwater. The EC values of all representative wells (or piezometers) are used for this purpose.

The groundwater salinity varies not only horizontally but also vertically; a zonation of groundwater salinity' is common in many areas. It is, therefore, advisable to prepare an EC map not only for the shallow groundwater but also for the deep groundwater. Other types of groundwater quality maps can be prepared by plotting different quality parameters (e.g. pH, Sodium Adsorption Ratio (SAR) values, etc.).

CHAPTER – 3

HYDRAULIC CONDUCTIVITY

3.1 HYDRAULIC CONDUCTIVITY

The hydraulic conductivity is a measure of the water-transmitting capacity of soils. It is the constant of proportionality in Darcy's law and is defined as the volume of water that will move through a porous medium in unit time under a unit hydraulic gradient through a unit area measured at right angles to the direction of flow. Hydraulic conductivity can have any units of length/time (e.g. m/day).

Bouwer and Jackson (1974) described in-situ methods for the determination of hydraulic conductivity (K-value). One method used to determine K above the water table and second used below the water table. Above the water table, the soil is not saturated. To measure the saturated hydraulic conductivity, one must therefore, apply sufficient water to obtain near-saturated conditions. These methods are called "infiltration methods" and use the relationship between the measured infiltration rate and hydraulic head to calculate the K-value. Below the water table, the soil is saturated. It then suffices to remove water from the soil, creating a sink and to observe the flow rate of the water into the sink together with the hydraulic head induced. These methods are called "extraction methods". The K-value can then be calculated with an equation selected to fit the boundary conditions. Camp (1977) determined hydraulic conductivity for an alluvial soil. Reynolds and Elrick (1985) used Guelph permeameter to find out In-situ saturated hydraulic conductivity and sorptivity.

Table 3.1. Hydraulic Conductivity of different Soil Types

S.N.	Soil type (texture)	Hydraulic conductivity, (m/day)
1	Dense clay (no cracks, pores)	< 0.002
2	Clay loam, clay (poorly structured)	0.002 - 0.2
3	Loam, clay loam, clay (well-structured)	0.5 - 2.0
4	Sandy loam, fine sand	1 - 3
5	Medium sand	1 - 5
6	Coarse sand	10 -50
7	Gravel	100 – 1000

Darcy's Law: The soil's hydraulic conductivity defined as the constant of proportionality in Darcy's Law:

$$V = K\, dh/dx$$

Where,

V = apparent velocity of the groundwater (m/d)

K = hydraulic conductivity (m/d)

h = hydraulic head (m)

x = distance in the direction of groundwater flow (m)

In Darcy's Equation, dh/dx represents the hydraulic gradient (s), which is the difference of h over a small difference of x. Hence, the hydraulic conductivity can be expressed as

$$K = V/s$$

and can thus be regarded as the apparent velocity (m/d) of the groundwater when the hydraulic gradient equals unity (s = 1).

In practice, the value of the hydraulic gradient is generally less than 0.1, so that v is usually less than 0.1K. Since the value of K is also usually less than 10 m/d, it follows that v is almost always less than 1 m/d.

The K-value of a saturated soil represents its average hydraulic conductivity, which depends mainly on the size, shape, and distribution of the pores. It also depends on the soil temperature and the viscosity and density of the water.

In unstructured sandy sediments soils, the hydraulic conductivity is the same in all directions. Usually, however, the value of K varies with the direction of flow. The vertical permeability of the soil or of a soil layer is often different from its horizontal permeability because of vertical differences in texture, structure, and porosity due to a layered deposition or horizon development and biological activity. A soil in which the hydraulic conductivity is direction-dependent is an-isotropic.

Anisotropy plays an important role in land drainage, because the flow of groundwater to the drains can, along its flow path, change from vertical to horizontal and intermediate between them. The hydraulic conductivity in horizontal direction is indicated by K_h, in vertical direction by K_v, and in an intermediate direction, especially in the case of radial flow to a drain, by K_r. The value of K_r for radial flow is often computed from the geometric, or logarithmic, mean of K_h and K_v (Boumans 1976):

$$K_r = (K_h\, K_v)^{1/2}$$

or

$$\log K_r = 0.5\, (\log K_h + \log K_v)$$

3.2 HYDRAULIC CONDUCTIVITY VARIABILITY

The hydraulic conductivity of a soil profile can be highly variable from place to place, and will also vary at different depths and within a soil layer. In alluvial soils, impermeable layers do not usually occur at shallow depth. In subsurface drainage systems in alluvial soils, not only the K-values at drain depth are important, but also the K-values of the deeper soil layers.

Soils with layers of low hydraulic conductivity or with impermeable layers at shallow depth are mostly associated with heavy, montmorillonitic or smectitic clay (vertisols), with illuviated clay in the sandy or silty layer at 0.5 to 0.8 m depth, or with an impermeable bedrock at shallow depth. Vertisols are characterized by a gradually decreasing K-value with depth because the topsoil is made more permeable by physical and biological processes, whereas the subsoil is not. Moreover, these soils are subject to swelling and shrinking upon wetting and drying, so that their K-value is also variable with the season, being smaller during the humid periods when drainage is required. If subsurface drains are to be installed, the K- values must be measured during the humid period. Seasonal variability studies are therefore important. If subsurface drainage of these soils is to be cost-effective, the drains must be installed at shallow depth (< 1 m).

Planosols can occur in tropical climates with high seasonal rainfalls. Under such conditions, a high drainage capacity is required and, if the impermeable layer is shallower than approximately 0.8 m, the cost-effectiveness of subsurface drainage becomes doubtful.

Hydraulic conductivity variability within soil layers

Measured K-values of a soil often show a log-normal distribution with a wide variation (Dieleman and Trafford, 1976). The plot of the logarithm of K-values against the cumulative frequency on normal probability paper. The figure shows that, except for the two highest and lowest observations, the data obey the log-normal distribution.

A representative value of K may be found from the geometric mean n

$$K^* = \sqrt[n]{(K_1 \times K_2 \times \ldots \times K_n)}$$

where n is the total number of observations

Taking the log value of K* we find from Equation that

$$\text{Log } K^* = (\log K_1 + \log K_1 + \ldots + \log K_1)/n$$

From this equation, we can see that log K* is the arithmetic mean of the log K-values. This corresponds to the mean value of the log-normal distribution.

The representative K-value of a soil layer can therefore be found simply as the modal or the median value of the frequency distribution of the observed K-values without log-transformation. Bouwer and Jackson (1974) conducted electric model tests with randomly distributed electric resistances to represent randomly varied K-values, and found that the geometric mean gave the most representative value.

Hydraulic conductivity variability between soil layers

When a soil shows a distinct layering, it is often found that the representative K-values of the layers differ. Generally, the more clayey layers have a lower K-value than the more sandy layers, but this is not always true. The representative value of K in layered soils depends on the direction of flow of the groundwater. When the water flows parallel to the soil layers, the representative value is based on a summation of the hydraulic transmissivities of the layers, but, when the water flows perpendicular to the layers, one uses a summation of the hydraulic resistances of the layers. The total transmissivity of soil layers for flow in the direction of the layers is calculated as

$$K^*D_t = \sum_{i=1}^{n} K_iD_i$$

where

K^* = weighted average K-value of the soil layers (m/d)

D_t = total thickness of the soil layers (m)

i = number of the soil layer

The value KiDi represents the hydraulic transmissivity for flow (m^2/d) of the i-th soil layer.

3.3. METHODS OF DETERMINATION OF HYDRAULIC CONDUCTIVITY

Correlation methods

The correlation methods for determining K-values in drainage surveys are frequently based on relationships between the K-value and one or more of the following soil properties: texture, pore-size distribution, grain-size distribution, or with the soil mapping unit.

Soil Texture

Soil texture refers to the per centage of sand, silt, and clay particles in the soil. Texture or textural class is often used for the correlation of K- values with other hydraulic properties of the soil (e.g. water-holding capacity an drainable pore space)

Pore-Size Distribution of the Soil

The pore-size distribution, the regularity of the pores, and their continuity has a great influence on the soil's K- values. Nevertheless, the study and characterization of the porosity aiming at an assessment of the K-values is not sufficiently advanced to be practical on a large scale. Applying Poiseuille's Law to a number of fractions of the pF-curve, he was able to calculate the K-value. Marshall's method is mainly applicable to granular (sandy) soils having no systematic continuous pores.

Grain-Size Distribution of the Soil

In sandy soils, which have no systematic continuous pores, the soil permeability is related to the grain-size distribution. Determining the K- value from the grain-size distribution uses the specific surface ratio (U) of the various grain-size classes. This U-ratio is defined as the total surface area of the soil particles per unit mass of soil, divided by the total surface area of a unit soil mass consisting of spherical particles of 1 cm diameter. The U-ratio, the porosity, and a shape factor for the particles and the voids allow us to calculate the hydraulic conductivity.

This method is seldom used in land drainage practice because the homogeneous, isotropic, purely-granular soils to which it applies are rare.

Small-scale *in-situ* methods

Bouwer and Jackson (1974) have described numerous small-scale in-situ methods for the determination of K-values. The methods fall into two groups: those that are used to determine K above the water table and those that are used below the water table. Above the water table, the soil is not saturated. To measure the saturated hydraulic conductivity, one must therefore apply sufficient water to obtain near-saturated conditions. These methods are called "infiltration methods" and use the relationship between the measured infiltration rate and hydraulic head to calculate the K-value. The equation describing the relationship has to be selected according to the boundary conditions induced.

Below the water table, the soil is saturated by definition. It then suffices to remove water from the soil, creating a sink, and to observe the flow rate of the water into the sink together with the hydraulic head induced. These methods are called "extraction methods". The K-value can then be calculated with an equation selected to fit the boundary conditions.

The small-scale in-situ methods are not applicable to great depths. Hence, their results are not representative for deep aquifers, unless it can be verified that the K-values measured at shallow depth are also indicative of those at greater depths and that the vertical K-values are not much different from the horizontal values. In general, the results of small-scale methods are more valuable in shallow aquifers than in deep aquifers.

Extraction Methods

The most frequently applied extraction method is the "auger-hole method". It uses the principles of unsteady- state flow. An extraction method based on steady-state flow has been presented by Zangar (1953) and is called the "pumped-borehole method". The "piezometer method" is based on the same principle as the auger-hole method, except that a small tube is inserted into the hole, leaving a cavity of limited height at the bottom. In sandy soils, the water-extraction methods may suffer from the problem of instability, whereby the hole caves in and the methods are not applicable. If filters are used to stabilize the hole, there is still the risk that sand will penetrate into the hole from below the filter, or that sand particles will block the filter, which makes the method invalid.

In clayey soils, on the other hand, where the K-value depends on the soil structure, it may happen that the augering of the hole results in the loss of structure around the wall. Even repeated measurements, whereby the hole is flushed several times, may not restore the structure, so that unrepresentatively low K-values are obtained (Bouma *et al.*, 1979).

As the depth of the hole made for water extraction is large compared to its radius, the flow of groundwater to the hole is mainly horizontal and one therefore measures a horizontal K-value. The water-extraction methods measure this value for a larger soil volume (0.1 to 0.3 m^3) than the laboratory methods that use soil cores. Nevertheless, the resulting variation in K-value from place to place can still be quite high. Using the auger-hole method, Davenport (Bentley *et al.*, 1989) found K-values ranging from 0.12 to 49 m/d in a 7 ha field with sandy loam soil.

Infiltration Methods

The "infiltration methods" can be divided into steady-state and unsteady- state methods. Steady-state methods are based on the continuous application of water so that the water level (below which the infiltration occurs) is maintained constant. One then awaits the time when the infiltration rate is also constant, which occurs when a large enough part of the soil around and below the place of measurement is saturated. An example of a steady-state infiltration method is the method of Zangar or "shallow well pump-in method" (e.g. Bouwer and Jackson, 1974).

Unsteady-state methods are based on observing the rate of draw down of the water level below which the infiltration occurs, after the application of water has been stopped. This measurement can start only after sufficient water has been applied to ensure the saturation of a large enough part of the soil around and below the place of measurement. Most infiltration methods use the unsteady-state principle, because it avoids the difficulty of ensuring steady-state conditions.

When the infiltration occurs through a cylinder driven into the soil, one speaks of "permeameter methods". Bouwer and Jackson (1974) presented a number of unsteady-state permeameter methods. They also discuss the "double-tube method", where a small permeameter is placed inside a large permeameter. The unsteady-state method whereby an uncased hole is used is called the "inversed auger-hole method". This method is similar to the Zangar and Guelph methods, except that the last two use the steady-state situation. In sandy soils, the infiltration methods suffer from the problem that the soil surface through which the water infiltrates may become clogged, so that too low K-values are obtained. In clayey soils, on the other hand, the infiltrating water may follow cracks, holes, and fissures in the soil, so that too high K-values are obtained.

In general, the infiltration methods measure the K-value in the vicinity of the infiltration surface. It is not easy to obtain K-values at greater depths in the soil. Depending on the dimensions of the infiltrating surface, the infiltration methods give either horizontal K-values (K_h), vertical K-values (K_v), or K-values in an intermediate direction. Although the soil volume over which one measures the K-value is larger than that of the soil cores used in the laboratory, it is still possible to find a large variation of K-values from place to place. A disadvantage of infiltration methods is that water has to be transported to the measuring site. The methods are therefore, more often used for specific research purposes than for routine measurements on a large scale.

3.4 THE AUGER-HOLE METHOD (SMALL-SCALE IN-SITU METHOD)

Principle

A hole is bored into the soil with an auger to a certain depth below the water table. When the water in the hole reaches equilibrium with the groundwater, part of it is removed. The groundwater then begins to seep into the hole and the rate at which it rises is measured. The hydraulic conductivity of the soil is computed with a formula or graph describing the mutual relationship between the rate of rise, the groundwater conditions, and the geometry of the hole. This method measures the average hydraulic conductivity of a soil column about 30 cm in radius and extending from the water table to about 20 cm below the bottom of the hole, or to a relatively impermeable layer if it occurs within 20 cm of the bottom.

Theory

As reported by van Beers (1970) and Bouwer and Jackson (1974), Ernst developed the following equation for the hydraulic conductivity of the soil in dependence of the average rate of rise of the water level in the hole.

$$K = C\frac{H_0 - H_t}{t}$$

Where,

K = hydraulic conductivity of the saturated soil (m/d)

C = a factor as defined in Equation below.

t = time elapsed since the first measurement of the level of the rising water in the hole (sec)

H_t = depth of the water level in the hole below reference level at time t (cm)

$H_o = H_t$ when t = O

The C-factor depends on the depth of an impermeable layer below the bottom of the hole (D) and the average depth of the water level in the hole below the water table (h') as follows:

When D> ½ D_2, then

$$C = \frac{4000\frac{r}{h'}}{\left(20+\frac{D_2}{r}\right)\left(2-\frac{h'}{D_2}\right)}$$

When D = 0, then

$$C = \frac{3600\frac{r}{h'}}{\left(10+\frac{D_2}{r}\right)\left(2-\frac{h'}{D_2}\right)}$$

where,

D = depth of the impermeable layer below the bottom of the hole (cm)

D_2 = depth of the bottom of the hole below the water table (cm), with the condition $20 < D_2 < 200$

r = radius of the hole (cm): $3 < r < 7$

h' = average depth of the water level in the hole below the water table (cm), with the condition: h' > D2/5

When $0 < D < \frac{1}{2} D_2$, one must interpolate between the results of the above two equations.

The value of h' can be calculated from

$$h' = 0.5 (Ho + H_n) - D_1$$

where

D_1 = depth of the water table below reference level (cm)

H_n = depth of the water level in the hole at the end of the measurements (cm)

Equipment and Procedure

The equipment consists of a tube, 60 cm long, the bottom end of which is fitted with a clack valve so that it can be used as a bailer. Extension pieces can be screwed to the top end of the tube. A float, a light-weight steel tape, and a standard are also part of the equipment. The standard is pressed into the soil down to a certain mark, so that the water-level readings can be taken at a fixed height above the ground surface.

The hole must be made with a minimum disturbance to the soil. The open blade auger used is very suitable for wet clay soils, whereas the closed pothole auger is excellent in dry soils.

The optimum depth of the holes depends on the nature, thickness, and sequence of soil layers, on the depth of the water table, and on the depth at which one wishes to determine the hydraulic conductivity. When augering the hole in slowly permeable soils, one often observes that the water is entering the hole only when the depth of the hole is well below the watertable. As the hole is deepened further, the water enters faster, because the rate of inflow of the water is governed by the difference between the water table and the water level in the hole, and by the depth of the hole below reference level (D_2). Sometimes, this phenomenon is incorrectly attributed to artesian pressure, but artesian pressure only exerts an influence if one pierces a completely or almost impermeable layer.

When the water in the hole is in equilibrium with the groundwater, the level is recorded. Water is then bailed out to lower the level in the hole by 20 to 40 cm. Measuring the rate of rise in the water level must begin immediately after bailing. Either the time for fixed intervals of rise, or the rise for fixed intervals of time can be recorded. The first technique requires the use of chronometers, while the second, which is customary in The Netherlands, needs only a watch with a good second hand. Normally, some five readings are taken, as these will give a reliable average value for the rate of rise and also provide a check against irregularities. The time interval at which water-level readings are taken is usually from 5 to 30 seconds, depending on the hydraulic conductivity of the soil, and should correspond to a rise of about 1 cm in the water level. A good rule of thumb is that the rate of rise in mm/s in an 8 cm diameter hole with a depth of 70 cm below the water table approximately equals the K-value of the soil in m/d.

Care should be taken to complete the measurements before 25% of the volume of water removed from the hole has been replaced by inflowing groundwater. After that, a considerable funnel-shaped water table develops around the top of the hole. This increases resistance to the flow around and into the hole. This effect is not accounted for in the formulas or flow charts developed for the auger-hole method and consequently it should be checked that $H_0 - H_t < 0.25 (H_0 - D_1)$.

After the readings have been taken, the reliability of the measurements should be checked. The difference in water level between two readings (ΔH) is therefore, computed to see whether the consecutive readings are reasonably consistent and whether the value of ΔH gradually decreases.

It often happens that ΔH is relatively large for the first reading, because of water dripping along the walls of the hole directly after bailing. Further inconsistencies in ΔH values may be caused by the float sticking to the wall or by the wind blowing the tape against the wall. Consistency can be improved by tapping the tape regularly. An example of recorded data and the ensuing calculations is presented in Table 2.

The auger-hole method measures the K-value mainly around the hole. It gives no information about vertical K-values nor about K-values in deeper soil layers. The method is therefore, more useful in shallow than in deep aquifers.

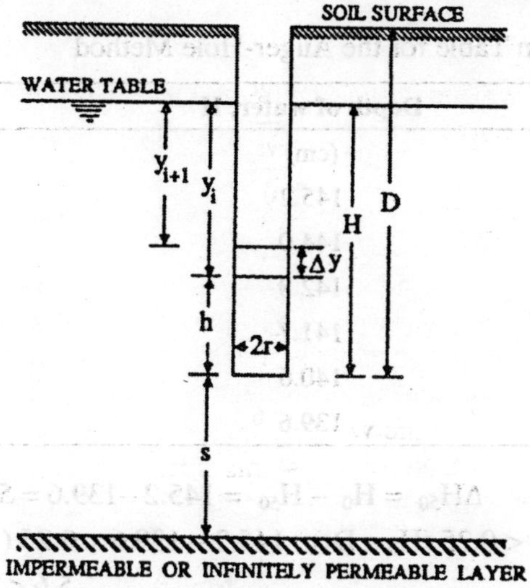

Figure 3.1. Auger-hole method

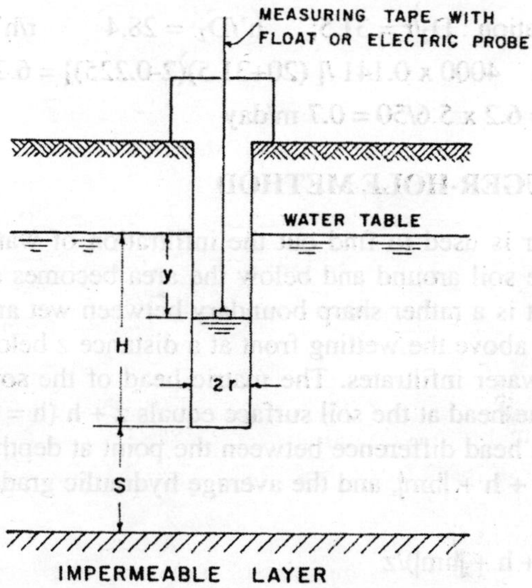

Figure 3.2. Measurement for the Auger-Hole Method

Example of measurements and calculations with the auger-hole method

No.: Location:
Date: Details:

Depth of auger-hole D' : 240 cm below reference
Depth of water table D1 : 114 cm below reference
D2 = D' – D1 : 126 cm
Auger-hole radius : 4 cm
Depth impermeable layer : D > ¼ D2

Table 3.2. Observation Table for the Auger-Hole Method

Time, t	Depth of water, H	ΔH *
(s)	(cm)	(cm)
0	145.2	-
10	144.0	1.2
20	142.8	1.2
30	141.7	1.1
40	140.6	1.1
50	139.6	1.0

Try t = 50 s; $\Delta H_{50} = H_0 - H_{50} = 145.2 - 139.6 = 5.6$ cm

Check $H_0 - H_{50} < 0.25\,(H_0 - D_1)$; $145.2 - 139.6 < 0.25\,(145.2 - 114)$;
$5.6 < 7.8 \rightarrow$ O.K. **)

Equation : $h' = 0.5\,(145.2 + 139.6) - 114 = 28.4$ cm

Ratio's for Equation : $D_2/r = 31.5$; $h'/D_2 = 28.4$ $r/h' = 0.141$

Equation : $C = 4000 \times 0.141 / [(20+31.5)(2-0.225)] = 6.2$

Equation : $K = 6.2 \times 5.6/50 = 0.7$ m/day

3.5. INVERSED AUGER-HOLE METHOD

An infiltrometer is used to find out the infiltration of water continuously into unsaturated soil. The soil around and below the area becomes almost saturated and that the wetting front is a rather sharp boundary between wet and dry soil. We shall consider a point just above the wetting front at a distance z below the soil surface in the area where the water infiltrates. The matric head of the soil at this point has a (small) value hm. The head at the soil surface equals z + h (h = height of water level in the cylinder). The head difference between the point at depth z and a point at the soil surface equals z + h + |hm|, and the average hydraulic gradient between the two points is:

$$S = (z + h + |hm|)/z$$

If z is large enough, s approximates unity. Hence, from Darcy's Law, we know that the mean flow velocity in the wetted soil below approaches the hydraulic conductivity (v = K), assuming the wetted soil is practically saturated.

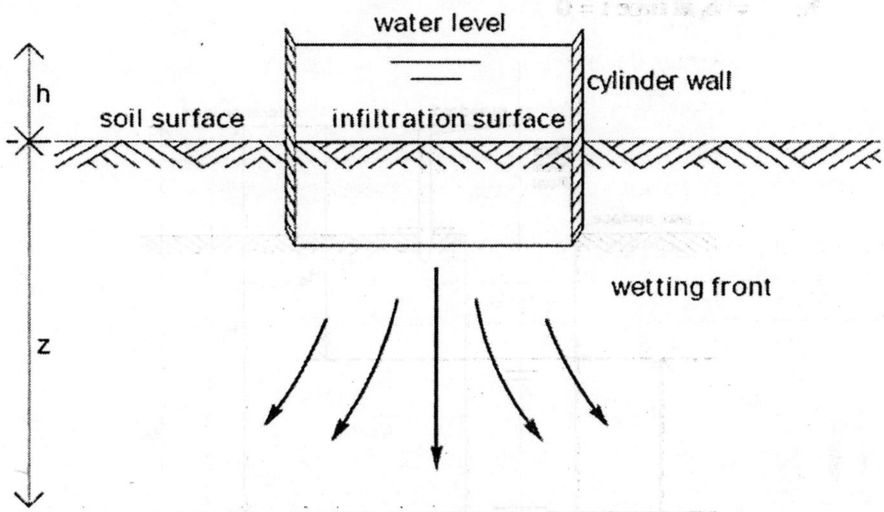

Figure 3.3. Infiltration process beneath a Cylinder Infiltrometer

If one bores a hole into the soil and fills this hole with water until the soil below and around the hole is practically saturated, the infiltration rate v will become more or less constant. The total infiltration Q will then be equal to v×A (where A is the surface area of infiltration).

With $v = K$,

we get: $Q = K \times A$.

For the inversed auger-hole method, infiltration occurs both through the bottom and the sidewalls of the hole.

Hence we have $A = \pi r2 + 2\pi rh$

(where r is the radius of the hole and h is the height of the water column in the hole).

So we can write $Q = 2\pi Kr\,(h + \frac{1}{2} r)$.

Further, we can find Q from the rate at which the water level in the hole is lowered:

$$Q = -\pi r2dh/dt.$$

Eliminating Q in both expressions gives

$$2K(h + \tfrac{1}{2} r) = r\,dh/dt.$$

Upon integration and rearrangement, we obtain

$$K = 1.15\,r\,\frac{\log\,(h_0 + \frac{1}{2}r) - \log\,(h_t + \frac{1}{2}r)}{(t - t_0)}$$

where

t = time since the start of measuring (s)

ht = the height of the water column in the hole at time t (cm)

h_0 = h_t at time t = 0

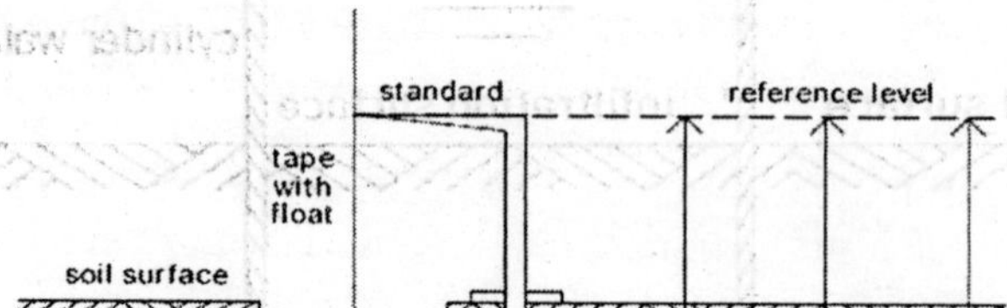

Figure 3.4. Measurements for the Inversed Auger-Hole Method

The values of h_t are obtained from

$$h_t = D' - H_t$$

where

D' = the depth of the hole below reference level (cm)

H_t = the depth of the water level in the hole below reference level (cm)

When H and t are measured at appropriate intervals, K can be calculated.

On semi log paper, plotting $h_t + \frac{1}{2} r$ on the log axis and t on the linear axis produces a straight line with a slope:

$$\tan\alpha = \frac{\log\left(h_0 + \frac{1}{2}r\right) - \log\left(h_t + \frac{1}{2}r\right)}{(t - t_0)}$$

The calculation of K with Equation can therefore, also be done with the value of tan α.

Hence: $K = 1.15\, r \tan\alpha$

Procedure

A hole is bored into the soil with an auger to a certain depth below the ground level. The hole must be made with a minimum disturbance to the soil. The optimum depth of the holes depends on the nature, thickness, and sequence of soil layers, on the depth of the water table, and on the depth at which one wishes to determine the hydraulic conductivity. After a hole is augered in the soil to the required depth, the hole is filled with water, which is left to drain away freely. The hole is refilled with water several times until the soil around the hole is saturated over a considerable distance and the infiltration (rate) has attained a more or less constant value. After the last refilling of the hole, the rate of drop of the water level in the hole is measured with the float and tape system. A float, a light-weight steel tape, and a standard are also part of the equipment. The standard is pressed into the soil down to a certain mark, so that the water-level readings can be taken at a fixed height above the ground surface. Calculate the $\left(ht+\frac{r}{2}\right)$ for all the time and take log of $\left(ht+\frac{r}{2}\right)$ of respective value. Plot $\left(ht+\frac{r}{2}\right)$ on log scale or log $\left(ht+\frac{r}{2}\right)$ on normal scale (y) and time t on linear scale (x). Draw a best fit straight line with observations. Calculate the slope of line from the plotted line.

Theoretical consideration

The hydraulic conductivity can be calculate by using following formula

$$K = 1.15 \times r \times \tan\alpha$$

Where,

$$\tan\alpha = \frac{\log\left(h_0+\frac{r}{2}\right)-\log\left(h_t+\frac{r}{2}\right)}{t-t_0}$$

K = hydraulic conductivity (cm/day)

h = The height of the water column in the in auger hole, cm.

h_o & h_t = the h value at time, cm,

h_o = h, at time t = 0.

t = time since the start of measuring (sec)

r = radius of the auger hole (cm)

tanα = slope of graph plotted between $ht+\frac{r}{2}$ vs t

Observations

D' = Total depth of the auger hole below reference level = _____ cm

r = Radius of augur hole = _______ cm

Table 3.3. Observation table for the Inversed Auger-Hole Method

S.N.	t (sec)	D'(cm)	h'(cm)	h_t=D'-h'	h_t+r/2	log(h_t+r/2)
1						
2						
3						
4						

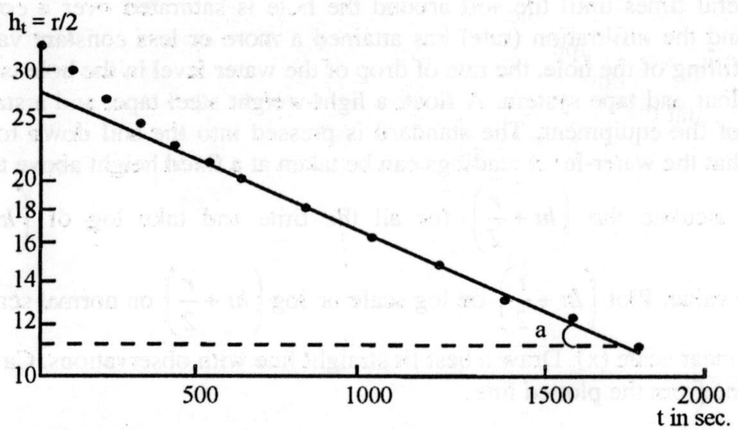

Figure 3.5. Fall of the water level, recorded with the inversed auger hole method, plotted against time (a).

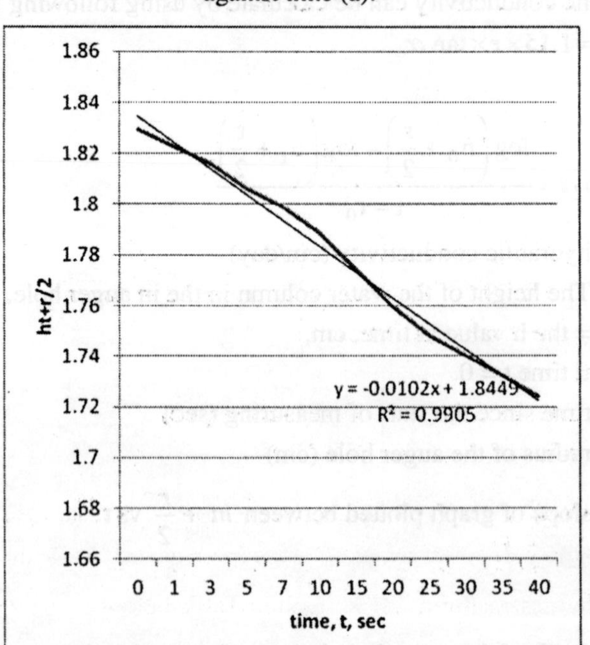

Figure 3.6. Fall of the water level, recorded with the inversed auger hole method, plotted against time (b).

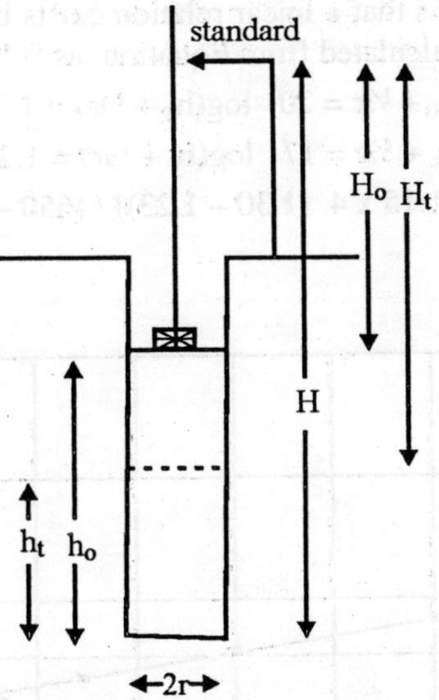

Figure 3.7. Infiltration form a water filled auger hole into the soil (Inversed auger hole method)

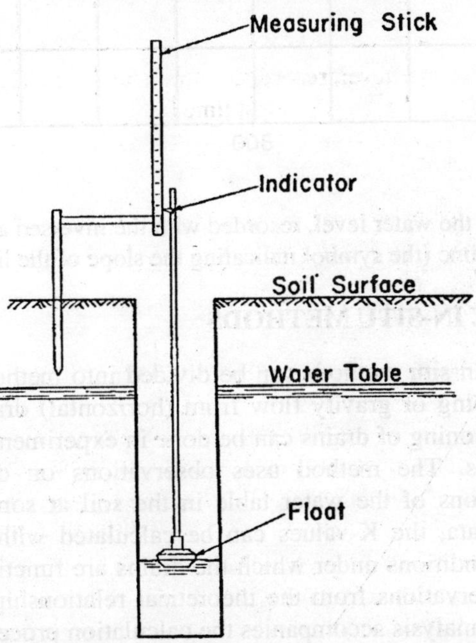

Figure 3.8. Measurements for the inversed auger hole method

An example of measurements is given in Table below.

Table. Example of measurements with inversed auger-hole method

(r = 4 cm, D' = 90 cm)

The data plotted shows that a linear relation exists between log (h_t + ½r) and t. The K-value can now be calculated from Equation as follows:

t_0= 140 h_0 + ½r = 20 log(h_0 + ½r) = 1.30

t = 650 h_t + ½r = 17 log(ht + ½r) = 1.23

K = [1.15 x 4 (1.30 – 1.23)] / [650 – 140]

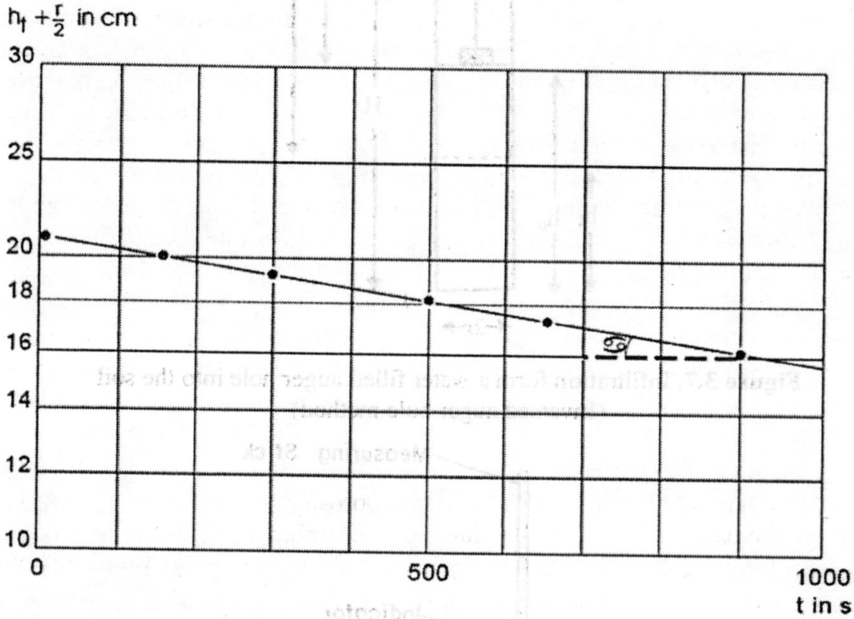

Figure 3.9. Fall of the water level, recorded with the inversed auger-hole method, plotted against time (the symbol indicating the slope of the line should be α)

3.6. LARGE-SCALE IN-SITU METHODS

The large-scale in-situ methods can be divided into methods that use pumping from wells and pumping or gravity flow from (horizontal) drains. Determining K-values from the functioning of drains can be done in experimental fields, pilot areas, or on existing drains. The method uses observations on drain discharges and corresponding elevations of the water table in the soil at some distance from the drains. From these data, the K-values can be calculated with a drainage formula appropriate for the conditions under which the drains are functioning. Since random deviations of the observations from the theoretical relationship frequently occur, a statistical confidence analysis accompanies the calculation procedure.

The advantage of the large-scale determinations is that the flow paths of the groundwater and the natural irregularities of the K-values along these paths are automatically taken into account in the overall K-value found with the method. It is then not necessary to determine the variations in the K-values from place to place, in horizontal and vertical direction, and the overall K-value found can be used directly as input into the drainage formulas. A second advantage is that the variation in K-values found is considerably less than those found with small- scale methods.

3.7 DRAINABLE POROSITY

Drainable porosity, *f*, defined as the ratio of the drained voids volume to the total volume of the voids plus the volume of the soil particles, is a necessary parameter in all equations that predict drawdown. During drainage it is not normally a constant, but is related, among other things, to the water table depth. Both the time of drawdown and the shape of the water table depend on the particular way in which drainable porosity is related to water table depth. When all the pore spaces in a soil volume are filled with water, it is saturated soil. Saturated soil is the soil in which all the pore spaces in a soil are filled up with water. If a saturated soil is allowed to drain under the action of gravity, the part of the water that is loosely held in the soil pores will drain out. The water is held within the soil pores due to capillary forces. If the force of surface tension is more than the gravity force, the water will not drain. The more fine textured a soil is, the more will be the force of surface tension and the less will be the volume of water drained. In coarser soils less will be the surface tension and more the volume of the drained water. The drainable pore space or drainable porosity is the volume of water that an unconfined aquifer releases from storage per unit surface area of aquifer per unit decline of the watertable. Small pores do not contribute to the drainable pore space because the retention forces in them are greater than the weight of water. The ratio of change in amount of soil water and corresponding change in the level of water table is said to be drainable porosity.

The average porosity value for a drawdown depth is used for calculations. This average value is called the equivalent drainable porosity, f_{eq}. Drainable porosity at any point in the soil is not solely a function of water table depth but is also determined by the time during which the drainage is permitted before the water table falls to a new position. The time required for a prescribed drawdown is affected by the size, depth, and spacing of the drains. The equivalent drainable porosity of the soil profile was determined using Equation

$$f_{n(h)} = \frac{V_n - V_{n-1}}{A\left(h_n - h_{n-1}\right)}$$

where

f = equivalent drainable porosity;

h = water table depth, m;

V = cumulative outflow volumes, m³;

A = area through water falls, m²; and

n = subscript showing the position of water table depth.

The entire drained soil profile was divided into slabs of 1 cm thick. The time required to lower the water table from initial depth, h_O, to 1 cm below was obtained from the time-hydraulic head relationship. The total volume of water drained during that period is the area below the drain discharge hydrograph for the same period, which was obtained by integrating the equation developed for time-drain discharge relationship for the same time period. The drainable porosity of the slab in the profile was obtained by dividing the total volume drained by the slab thickness. The drainable porosities for successive slabs of 1 cm thickness were found out for the entire range of water table drawdown for all the drains. Relationships were then developed for drainable porosity with respect to the mid-spacing water table.

Drainable porosity (f) as a function of the water table depth or suction (h) can be measured using the apparatus (Heins' apparatus or Hanging Column apparatus).

PROCEDURE FOR DETERMINATION OF DRAINABLE POROSITY IN LABORATORY

- The sintered glass porous plate of the apparatus is glued to the funnel and should have a "Leak Value" (the pressure at which the water in the saturated plate is driven out and air starts filling the pores) at least equivalent to 5 m of water.
- Flexible transparent polythene tube (which permits detecting trapped air bubbles), is fitted making sure that no air bubble is trapped below the bottom of the porous plate or anywhere in the water column in the tube.
- A burette or graduated glass tube (to read volume of water directly) connection is made as shown and the water level in it is maintained at the same level as the top of the porous plate.
- Undisturbed soil sample is taken from the field in a metal ring of known internal volume and is placed on the porous plate after making the two surfaces of the sample flush with the edges of the ring. This ensures a firm contact between the porous plate and the sample.
- The soil in the sample is saturated by maintaining the water column in the burette slightly above the top of the porous plate. Saturation may take 24 hours.
- The burette reading is noted, excess water (if any) on the top of the porous plate is soaked in a blotting paper.
- The top of the funnel is closed by a plastic sheet to avoid evaporation loss from the sample during the test.
- A pin hole is made in the plastic sheet to maintain atmospheric pressure over the sample, if the experiment is conducted in the dry season, it is necessary to place a moist sponge over the pin hole to minimize evaporation loss from the sample.
- The burette is lowered by a certain distance (usually in steps of 10 cm at a time).
- The lowering of the burette imparts a suction to the porous plate and to the saturated soil sample, which starts draining water slowly.
- The drained water raises the water level in the burette. At certain suction, the rise will be more which will require readjustment of the burette height such that the originally applied suction is closely maintained.
- When there is no further rise of water column in the burette, the elevation difference between the top of the porous plate and the water surface in the burette are noted, which is the average suction applied to the sample.

- The difference between the initial and the final burette readings, gives the volume of water drained from the sample consequent to the applied suction.
- This process is repeated by lowering the burette in steps of 10 cm initially and by more extent later till the desired suction, which may be the maximum possible depth of the subsurface drain, or ascertained based on any other criteria.
- Based on the cumulative volume of water drained, drainable porosity is calculated by dividing the volume of water drained by the volume of the sample.

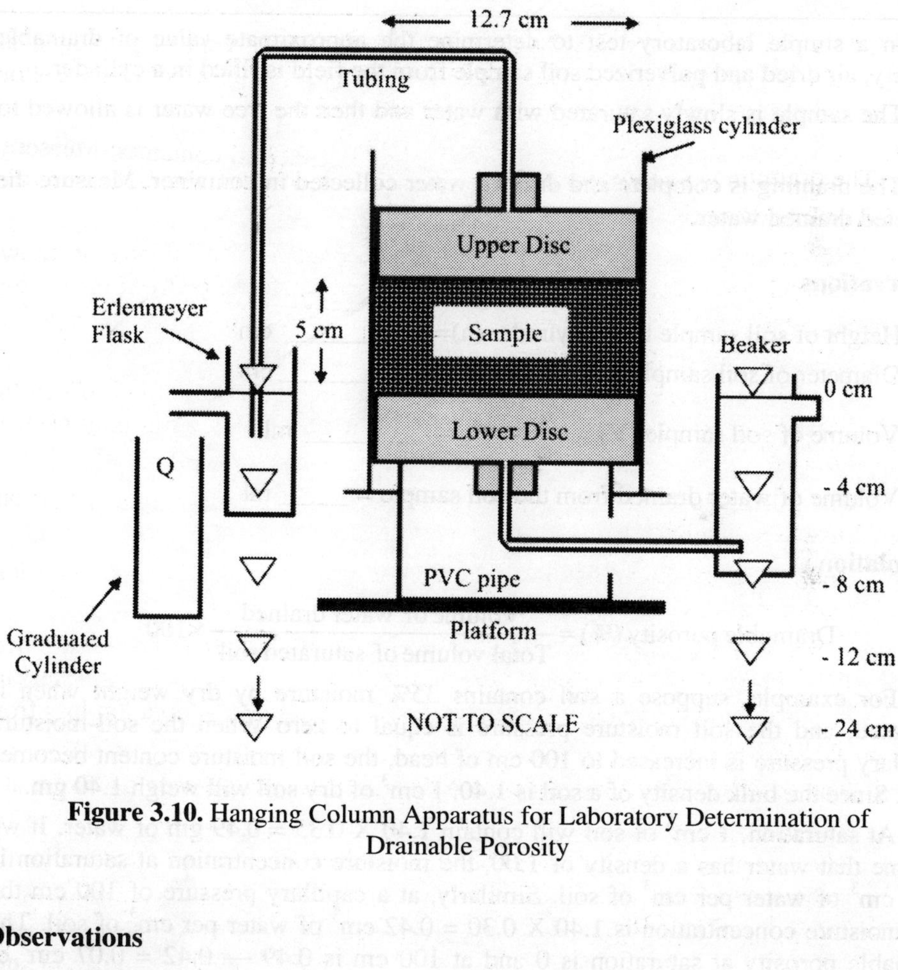

Figure 3.10. Hanging Column Apparatus for Laboratory Determination of Drainable Porosity

Observations

Diameter of ring = ______________ cm

Height of ring = ______________ cm

Internal volume of ring = ______________ cubic cm

Table 3.4. Observation table for the Inversed Auger-Hole Method

Sr. No.	Burette water surface elevation		Burette reading, ml		Suction of water column, cm	Volume of water extracted, ml		Drainable porosity, fraction
	Initial	Final	Initial	Final		Incremental	Cumelative	
1								
1								
2								
3								

In a simple laboratory test to determine the approximate value of drainable porosity, air dried and pulverized soil sample from the field is filled in a cylinder.

The sample is slowly saturated with water and then the free water is allowed to drain.

The draining is complete and drained water collected in container. Measure the collected drained water.

Observations

Height of soil sample in the cylinder (h)= ________ cm

Diameter of soil sample in the cylinder (d) = ________ cm

Volume of soil sample (V) = $\frac{\pi}{4} d^2 * h =$ ________ ml

Volume of water drained from the soil sample = ______ ml

Calculation

$$\text{Drainable porosity (\%)} = \frac{\text{Volume of water drained}}{\text{Total volume of saturated soil}} \times 100$$

For example, suppose a soil contains 35% moisture by dry weight when it saturated, and the soil moisture pressure is equal to zero When the soil-moisture capillary pressure is increased to 100 cm of head, the soil moisture content becomes 30%. Since the bulk density of a soil is 1.40, 1 cm^3 of dry soil will weigh 1.40 gm.

At saturation, 1 cm^3 of soil will contain 1.40 X 0.35 = 0.49 gm of water. If we assume that water has a density of 1.00, the moisture concentration at saturation is 0.49 cm^3 of water per cm^3 of soil. Similarly, at a capillary pressure of 100 cm the soil-moisture concentration is 1.40 X 0.30 = 0.42 cm^3 of water per cm^3 of soil. The drainable porosity at saturation is 0 and at 100 cm is 0.49 — 0.42 = 0.07 cm^3 of water per cm^3 of soil. This is the amount of water that will drain out of 1 cm^3 of soil at the soil surface when the watertable drops from the soil surface to a point 100 cm below the soil surface.

Rainfall/irrigation and water table rise

The approximate field method of determining drainable porosity is to note the water table rise on account of a known depth of rainfall or irrigation and dividing the latter by the former after expressing both in consistent units. Thus, if an irrigation application (or rainfall) of 5 cm causes a temporary water table rise of 25 cm, the drainable porosity is (5/25) = 0.2 or 20 per cent. Such a result will be inaccurate because, the runoff loss, the water needed to first bring the soil to field capacity and the natural internal drainage outflow between the time of irrigation or rainfall and the time of noting water table depth have not been considered and the entire amount of applied water has been assumed to have caused water table rise.

3.8 MEASUREMENT OF HYDRAULIC CONDUCTIVITY FALLING HEAD PERMEAMETER APPARATUS

The coefficient of permeability is defined as the rate of flow of water through a unit cross-sectional area normal to the direction of flow under a unit hydraulic gradient. It has the unit of velocity, *i.e.*, cm per hour, meters per day etc. It depends on the properties of fluid and porous medium. Permeability is a measure of the capacity of a soil to permit the passage of water and has the dimensions of velocity. Wit (1967) used the apparatus for measuring hydraulic conductivity of undisturbed soil samples in the laboratory. It is defined as the ability of soil to transmit the water. It is also defined as coefficient 'k' in Darcy's Law;

$$v = k\ i$$

Where,

v = velocity of seepage, m/sec,

i = hydraulic gradient

List of apparatus and materials

Soil sample, distiller water, permeameter apparatus, stop watch, scale, measuring jar, balance, moisture boxes, trimming knife etc.

Procedure

- Collect the undisturbed soil sample from the field using the dynamic ramming tool provided in the apparatus at a designed depth and thoroughly mix with distilled water to raise coefficient of permeability.
- Transfer the soil specimen into the peameameter mould using sampler extruder.
- Saturate the soil by keeping the mould in the water for nearly 24 hours by opening the air release valve.
- After saturation of soil specimen, place the mould assembly in the bottom tank and fill the tank with water up to its outlet.
- Make a coarser filter screen at the upper end and lower end of the sample.

- Connect the water inlet nozzle of the mould to the stand pipe filter with water; allow the water to seep down through the soil sample.
- Note the time interval required for the fall of water level to some final value from the initial value.
- Repeat the last step twice for determining the time against height of fall of water in the stand pipe.
- Perform a series of test using different sizes of soil sample, for determining the average value of hydraulic conductivity.

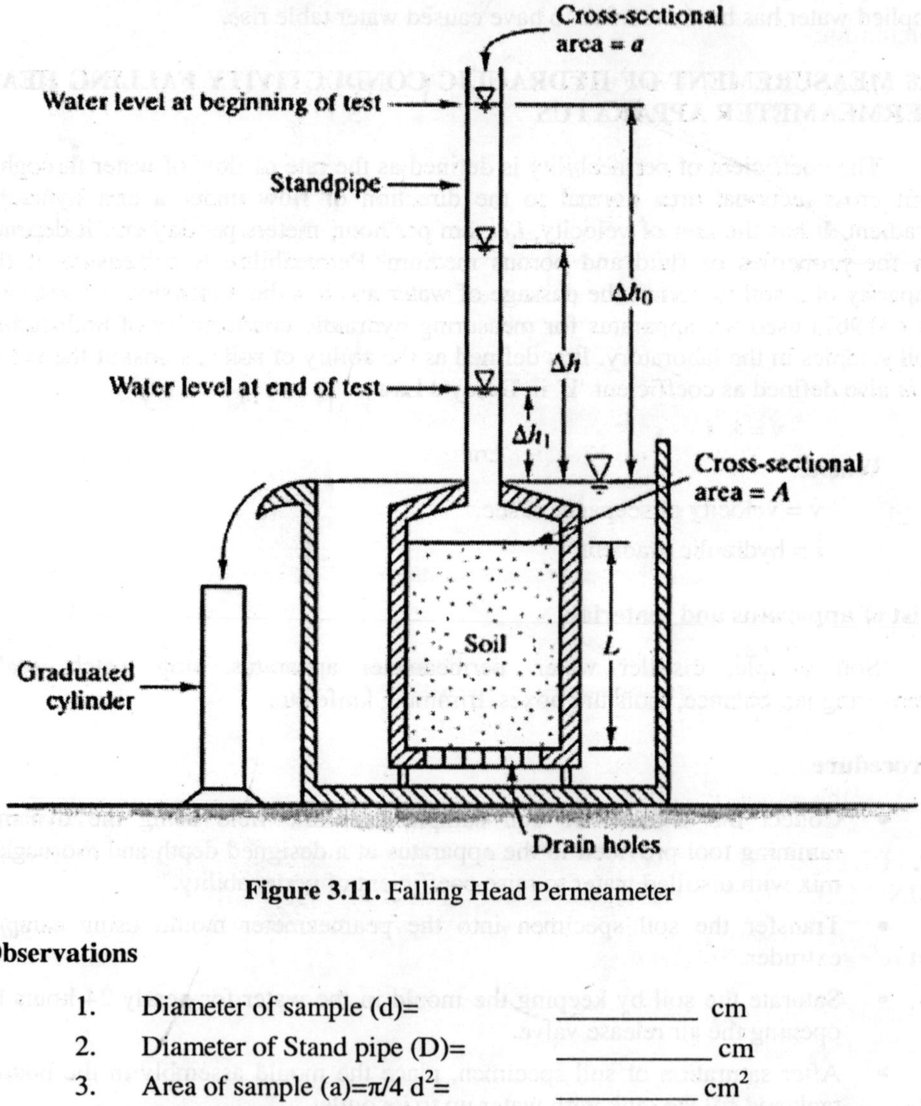

Figure 3.11. Falling Head Permeameter

Observations

1. Diameter of sample (d)= ____________ cm
2. Diameter of Stand pipe (D)= ____________ cm
3. Area of sample (a)=$\pi/4\ d^2$= ____________ cm^2
4. Area of stand pipe (A)= $\pi/4\ D^2$ = ____________ cm^2

Table 3.5. Observation table for Falling Head Permeameter

Sr. no	Sample length, L (cm)	Time, (sec)		Water level in the stand pipe, (cm)	
		t_1	t_2	Initial (h_1)	Final (h_2)
(1)	(2)	(3)	(4)	(5)	(6)

Calculation

The coefficient of permeability is determined as under.

$$K = \frac{a \times L}{A(t_2 - t_2)} ln\left(\frac{h_1}{h_2}\right) \quad \text{cm/day}$$

Apply a correction to the calculated value of coefficient of permeability to account for the effect of temperature on the viscosity of the permanent fluid as

Permeability at 27 ºC $K_{27} = K n / n_{27}$

Where,

K = permeability at test temperature

n = viscosity at test temperature

n_{27} = viscosity at 27 ºC temperature

Result

- The value of coefficient of permeability was found to be ______ cm/sec.

Precautions

- Soil sample should not be disturbed while transferring into the permeameter mould.
- Ensure that dynamic ramming tool makes contact with full surface area of the sample during ramming.

3.9. DETERMINATION OF THE COEFFICIENT OF PERMEABILITY USING CONSTANT HEAD PERMEAMETER

List of apparatus and materials

Soil sample, distilled water permeameter apparatus stop watch, scale, measuring jars, balance moisture boxes, trimming knife, etc.

Principle

If the tank is initially completely full without and air space, then as soon as water starts flowing out of the tank and air starts entering through the air intake tube, the pressure at the level of the bottom end of the tube becomes atmospheric (constant) and it remains so as long as the water level remains above the bottom end of the tube. The constant hydraulic head causing the flows through the elevations of the free water level in the bottom tank of the permeameter and the bottom end of the air intake tube. The head can be increased or decreased by lowering or raising the air intake tube.

Procedure

1. Collect the undisturbed soil sample from the field using the samples and the dynamic ramming tool provided in the apparatus at a desired depth and mix with distilled water thoroughly to raise the water content of the soil to saturate.
2. Transfer the soil specimen into the permeameter mould using sampler extruder.
3. Saturate the soil by keeping the permeameter mould in the water for nearly 24 hours by opening the air release valve.
4. After saturation of soil specimen, place the permeameter mould assembly in the bottom tank and fill the tank with water upto its outlet.
5. Place a coarser filter screen at the upper end and lowed end of the sample.
6. Connect the water inlet nozzle of the mould to the stand pipe filler with water.
7. Adjust the hydraulic head in the stand pipe containing water by raising or lowering the air intake tube within it.
8. Start the stop watch, allow the water to seep down through the sample and record the volume of water passing through the sample
9. Repeat the test twice or thrice under the same head and for the same time interval.

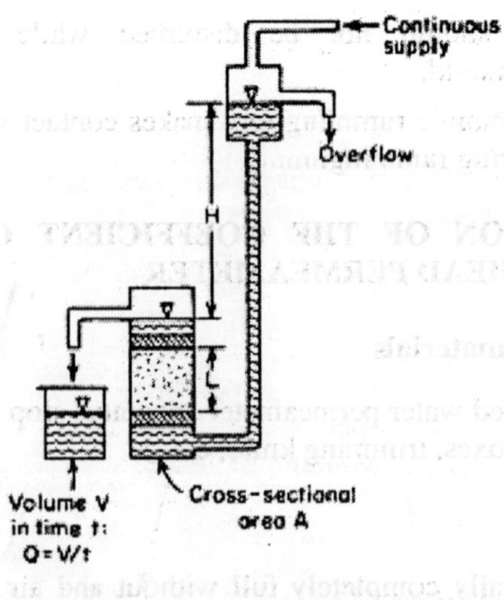

Figure 3.12. Constant Head Permeameter

Observations

(1) Hydraulic head (h) = ____________ cm

(2) Length of sample (L) = ____________ cm

(3) Cross sectional area of the sample (A) = ________ cm^2

(4) Time interval (t) = ________ sec.

(5) Temperature of the room = ____________ °C

(6) Quantity of water collected (Q)

(a) 1st test = ______________ ml

(b) 2nd test = ______________ ml

(c) 3rd test = ______________ ml

Table 3.6. Observation table for Constant Head Permeameter

Test No.	Sample length, L (cm)	Time, (sec) t_1	Time, (sec) t_2	Quantity of water collected (Q), ml
(1)	(2)	(3)	(4)	(5)

Calculations

The coefficient of permeability at test temperature is determined as:

$$K = Q L / A h t$$

Precautions

1. Soil sample should not be disturbed while transferring into the permeameter mould.
2. Ensure that dynamic ramming tool makes contact with full surface area of the sample during ramming.
3. Care must be taken while maintaining the head by lowering or raising the air intake pipe
4. Observations should be taken after ensuring that steady state condition of seepage exists.
5. Care must be taken to ensure that observations should not be recorded unless the water level is very near the bottom of the tube.

Remarks

(1) Reproduction of arrangement and distribution of particles, within the soil and density of the soil on site can hardly ever be achieved.

(2) Behavior of the mass of soil at the field hardly ever corresponds to the behaviour of small sample.

(3) The use of water other than site water may affect the nature of adsorbed water surrounding fine grained soil particles and may even lead to the chemical action.

(4) When the direction of seepage is horizontal and inclined vertical, value of permeability is of little practical value.

(5) k varies with void ratio, hence the levels of total pressure, pore water pressure and effective pressure values should be equivalent to the field values.

Exercises

1. A soil sample was taken from a field in a sampling ring of 12 cm diameter and 6 cm height. Another sample is taken in a soil auger for moisture content determination, which was 22 per cent on dry mass basis. The dry bulk density was found out for the sample as 1.3 g/cm^3. In the laboratory, 5 ml of water was needed to saturate the sample in the ring. Find the porosity of the sample and discuss the drain ability of the soil.
2. A soil has saturation moisture content of 37 per cent, field capacity moisture content of 20 per cent and a dry bulk density of 1.25 g/cm^3. Find the total porosity, the drainable porosity and discuss the drainage behaviour of the soil.
3. At a certain rime, the bulk density of the top soil layer was 1.25 g/crn^3. After five years of cultivation, the bulk density increased to 1.4g/cm^3. Calculate the change in the porosity of the soil if the particle density is 2.7 g/cm^3. Discuss the probable change in the drainage behaviour of the soil.
4. Assuming an average sand particle size 1 mm diameter, find the number of particles of 0.01 and 0.001 mm diameter to constitute the same mass as that of one sand particle. Hence calculate the specific surface.
5. Find the volume of water that will be retained in a saturated soil column of 1 cm^2 area and 1 m height, if the water table drops from surface to 60 cm below and find the corresponding moisture content. The average drainable porosity of a soil is 0.15. The bulk density of the soil is 1.4 g/cm^3 and the soil parade density is 2.7 g/cm^3.
6. The sample of soil from the field is compacted in a cylindrical container of 15 cm diameter and 30 cm height. The void ratio was attained as 0.7. The compacted soil is then slowly saturated and subsequendy, the water table is adjusted at a depth of 15 cm from the top. After a while, a total of 60 ml of water was drained out. Find the saturation moisture content, the moisture content after 60 ml of water has drained out and the average drainable porosity of the compacted sample. The soil particle density is 2.6 g/cm^3.
7. A rectangular container of size 3 cm by 5 cm and 15 cm high was filled with air dried and pulverized soil sample from the field to determine the approximate value of drainable porosity in the simple laboratory test. The container was tapped lightly from time to time while filling the soil sample to

achieve a stable particle arrangement. The sample was slowly saturated with water and then the free water was allowed to drain. The draining was complete in 4 hours and 35 ml of water drained. Find the drainable porosity.

8. Total 10 lateral drains laid at a spacing of 40 m and each 150 m long in a subsurface drainage network, join a collector drain. The average discharge at the outlet of the collector drain was 15 lps when the water table dropped from surface to 50 cm below surface in 3 days. Find the average drainable porosity of the soil.

9. Find the total porosity, the drainable porosity and discuss the drainage behaviour of the soil. The saturated moisture content of the soil was 35 per cent, field capacity moisture content of 18 per cent (both on dry mass basis) and a dry bulk density of 1.8 g/cc.

10. The bulk density of the 20 cm soil layer from the surface was 1.4 g/cc. After five years of cultivation, the bulk density increased to 1.8 g/cc. Calculate the change in the porosity of the soil if the particle density is 2.4 g/cm^3. Discuss the probable change in the drainage behaviour of the soil.

11. The average drainable porosity of a soil is 0.15. Find the volume of water that will be retained in a saturated soil column of 1 cm^2 area and 100 cm height if the water table drops from surface to 60 below and find the corresponding moisture content. The bulk density of the soil is 1.4 g/cm^3 and the soil parade density is 2.7 g/cc.

12. The void ratio of the soil sample was found out as 0.8 by compaction of the soil in a cylindrical container of 10 cm diameter and 20 cm height. The compacted soil is then slowly saturated and subsequently, the water table is adjusted at a depth of 20 cm from the top. After a while, a total of 50 ml of water was drained out. Find the saturation moisture content, the moisture content after 50 ml of water has drained out and the average drainable porosity of the compacted sample. The soil particle density is 2.6 g/cc.

13. Find the drainable porosity. In a laboratory test to determine the approximate value of drainable porosity, air dried and pulverized soil sample from the field is filled in a rectangular cross-section container 4 cm by 5 cm and 16 cm high (internal dimensions). The container was tapped lightly from time to time while filling the soil sample to achieve a stable particle arrangement. The sample was slowly saturated with water and then the free water was allowed to drain. The draining was complete in 4 hours and 45 ml of water drained.

14. The average discharge at the outlet of the collector drain was 11 lps when the water table dropped from surface to 40 cm below surface in 3 days. In a subsurface drainage network, 10 lateral drains laid at a spacing of 40 m and each 150 m long, join a collector drain. Find the average drainable porosity of the soil.

CHAPTER - 4

SURFACE DRAINAGE SYSTEM AND DESIGN

Surface drainage is the removal or diversion of excess water from the surface of land by means of improved natural or constructed channels to the outlet or natural drain. Drains may be either interceptor drain or relief drains. They may be either open drain or tile drain. An open ditch drainage system consists of a network of channels draining a given watershed. It consists of field ditches which collect surface water directly and empty into outlet ditches. The outlet ditches collect water from field ditches or tile drains and also directly from the land surface and empty into main drain.

Table 4.1. Comparison between Interceptor Drain and Relief Drain

Sr. no.	Interceptor drain	Relief drain
1.	Located almost perpendicular to the direction of ground-water flow.	Located almost parallel to the direction of ground-water flow.
2.	To intercept the seepage and runoff and drain it away.	To lower the water table.
3.	Water enters directly from the surface or by the seepage from the surrounding soil mass.	Water seeps into it from either side of the drain.
4.	It provides least interference to the farming operations.	It interfere the movement of machinery.
5.	Located along the foothill where the seepage from hillsides enters a flat land.	Located along the low lying area.
6.	It tends to isolate the problem area from the source of excess water.	It cut away the farm area into fragments, waste the farm area.

Note = Construction of the drain is always started at the outlet and progresses upstream.

The selection of surface drainage depends on topography, soil characteristics, crops and availability of suitable outlets.

4.1. AREA WHICH REQUIRES SURFACE DRAINAGE

The area which requires surface drainage are:

- Adequate outlets for tile drains are not available.
- Intensity rains occurs for short time and large quantity of water are to be drained quickly.

- Subsurface drainage is not practicable or economical, very slowly permeable soil such as heavy clay and shallow soils of 20 to 50 cm depth, with underlying very permeable subsoil is there.
- Slow infiltration, low permeability, or restricting layers in the profile prevent the ready absorption of high-intensity rainfall.

4.2. LAND FORMATION FOR SURFACE DRAINAGE

A surface drainage system always has two components: Open field drains to collect the ponding water and divert it to the collector drain. Land forming to enhance the flow of water towards the field drains: Land forming means changing the surface of the land to meet the requirements of surface drainage or irrigation. There are three land-forming systems: bedding, land grading and land planing.

Bedding

Bedding is the oldest surface drainage practice. With this system, the land surface is formed into beds. This work can be done by manual labour, animal traction, or farm tractors. The beds are separated by parallel shallow, open field drains, oriented in the direction of the greatest land slope. The water drains from the beds into the field drains, which discharge into a collector drain constructed at the lower end of the field and at right angles to the field drains. Bedding is a surface drainage method achieved by ploughing land to form a series of low beds, separated by parallel field drains. The bedding system is normally used for grassland. In modern farming, bedding is not considered an acceptable drainage practice for row crops, because rows near the field drains will not drain satisfactorily. To overcome the disadvantages of the bedding system, the two other methods of land forming have been developed: land grading and land planing.

Design considerations of Bedding system

To ensure good drainage in a bedding system, the beds should not be more than 10 m wide. Further, the width of the beds is governed by the following:

- The kind of crops to be grown: Field crops require narrower beds than permanent pasture or hay crops do.
- Farming operations on beds: Ploughing, planting, and cultivating should fit the width of a bed. Bed width should be a multiple of the effective width of farm equipment.
- Soil characteristics: Soils with low infiltration and low hydraulic conductivity require narrower beds than soils with better characteristics.

Construction of bedding

It often takes several years of ploughing to obtain an adequate bedding system. During the first ploughing, care should be taken to make beds of uniform width throughout the field and to have the field drains running in the direction of the greatest slope. Any obstructions or low points in the field drains should be eliminated because they will cause standing water and loss of crops. The collector drain should be laid out in the direction of the lesser field slope, and should be properly graded towards the main drainage system.

Land Grading and Land Planing

When grading land for surface drainage, the slope does not need to be made uniform, as for irrigation; a non-uniform slope will suffice.

In addition, the types of crop and how they will be grown have to be considered. Crops like maize, potatoes, and sugar cane are grown in rows with small furrows in between. For such crops, the length of the rows and the slopes of the field must be selected so as to avoid erosion and overtopping of the small furrows. To prevent erosion, it is recommended that the flow velocities in the furrows should not exceed 0.5 m/s. In highly erodible soils, the row length is limited to about 150 m. Slightly erodible soils allow longer rows, up to 300 m. The direction of the rows and furrows need not necessarily be at right angles to the slope, but can be selected in any way that meets the above recommendations.

Small grains and hay crops are grown by broadcast sowing or in rows, but on an even surface (i.e. no furrows). For such crops, surface drainage takes place by sheet flow. This flow is always in the direction of the maximum slope. With sheet flow, the flow resistance is much higher than in small furrows, and the flow velocity on the same land slope is less. Even after careful land grading and smoothing, however, sheet flow always has a tendency to concentrate in shallow depressions, and gullies are easily formed. With the transport duration for low flow velocities in mind, it is recommended that the field length in the flow direction be limited to 200 m or less.

For wet-land rice and other crops grown in basins, the surface is levelled by earthmoving machinery (large basins) or with simple farm implements. Levelled fields are surrounded by field bunds. Any excess water from basins is usually drained through an overflow in the field bunds that spills the water directly into a field drain.

Land grading

Land grading for surface drainage consists of forming the land surface by cutting, filling and smoothing it to predetermined grades, so that each row or surface slopes to a field drain. It is a one-time operation.

Land grading for surface drainage differs from land levelling for irrigation in that, for drainage, the grades need not be uniform. They can be varied as much as is needed to provide drainage with the least amount of earthmoving. Land grading is forming the surface of the land to predetermined grades, so that each row or surface slopes to a field drain

Compared with bedding, land grading reduces the number of field drains, thus reducing the need for weed control and maintenance. Land grading also means that more land is available for use.

Construction for land grading

Land grading can be done by the farmers, although normal farm equipment, even if mechanized, can handle small-scale grading operations or the maintenance of already established grades. Large-scale land grading is done by contractors with conventional earthmoving equipment or with laser-guided motorized graders.

Grading operations involve a number of steps. The first step is to prepare the site. If the land has already been cleared, the work mainly involves removing or destroying vegetation and other obstacles, and levelling ridges or rows. This can normally be done with farm equipment. The surface should be dry, firm, and well-pulverized to enable the equipment to operate efficiently

The second step is rough grading. This can be done with various types of equipment (e.g. dozers, motor graders, scrapers). The choice will depend on the soil conditions, the amount of earthwork needed, the time and equipment available, the size of the fields to be graded as one unit, and local experience.

The third step is the finished grading. On small fields, drags, harrows, and floats can be used. These implements can be pulled by a farm tractor or by animal traction. On larger fields, a land plane (a bottomless scraper) pulled by a farm tractor is used. For the final smoothing, several passes are usually made at angles to one another. When extensive grading is done with heavy equipment, it is likely to cause the soil to become compacted. This compaction should be relieved to eliminate differences in soil productivity. Various tillage tools can be used for this work (e.g. subsoilers, chisels, scarifiers, and rippers).

Land planing

Land planing is the process of smoothing the land surface to eliminate minor depressions and irregularities, but without changing the general topography. It is often done after land grading, because irregular micro-topography in a flat landscape, in combination with heavy soils, can cause severe crop losses.

Land planing is smoothing the land surface with a land plane to eliminate minor depressions and irregularities without changing the general topography. In the field, surface drainage systems can have two different layouts: the random field drainage system, and the parallel field drainage system.

4.3. SURFACE DRAINAGE SYSTEMS

Random field drainage system

The random field drainage system is applied where there are a number of large but shallow depressions in a field, but where a complete land-forming operation is not considered necessary. The random field drainage system connects the depressions by means of a field drain and evacuates the water into a collector drain. The system is often applied on land which does not require intensive farming operations (e.g. pasture land) or where mechanization is done with small equipment.

Parallel field drainage system

The parallel field drainage system, in combination with proper land forming, is the most effective method of surface drainage. The parallel field drains collect the surface runoff and discharge it into the collector drain. The spacing between the field drains depends on the size of fields that can be prepared and harvested economically, on the tolerance of crops to ponding, and on the amount and costs of land forming. The system is suitable in flat areas with an irregular micro-topography and where farming operations require fields with regular shapes.

4.4. SURFACE DRAINS

Geomorphology is a field of study that includes the study of natural drainage patterns. Some of the important surface drainage patterns found in nature are: *Annular, Trellis, Dichotomy. Parallel, Radial and Rectangular*. However, the artificial surface drainage systems can be classified as: *Random. Interception, Parallel or Diversion, Bedding and Field Drain Systems.* An engineered surface drainage system may include one or more of these types, and may incorporate segments of natural systems. The layout would depend on the topography, type of soil and the nature of crops grown.

Designing of a surface drain system will include the following aspects :

(i) its alignment - (plan form),

(ii) the longitudinal profiling,

(iii) capacity requirement,

(iv) design velocity, and

(v) the cross-sections along the drain length including side slope and bank width.

Various considerations go into fixing these features, vis-a-vis, a surface drainage system.

Components of Surface drainage system

1. Relief drain/Open ditch/ Interceptor drain
2. Collector drain / link drain
3. Sub main drain / main drain
4. Outlet

Alignment of Drains

While proposing the alignment of a drain following factors need be considered:

(a) The drain channel is preferably aligned along the natural drainage line, i.e., the valley line representing the lowest contour in the area,

(b) The alignment should be adjacent to ponds and marshy lands which are then connected to the drain, taking care to avoid aligning the drain through the ponds and marshy lands to the extent possible. This scheme would result random drain system,

(c) The drain alignment should avoid crossing irrigation channels which usually follow the ridge line in an area. Also, the number pf crossings with road and railways should be minimized, and

(d) In the case of long winding, meandering alignment of the natural drain it may be desirable to cut straight segments to reduce the drain length, while avoiding the encroachment into valuable cultivated land.

Longitudinal Profile of the drains

It is customary, in engineering practice, to represent distances along a drain (as along any line of interest) in terms of reduced distance (RD) starting from the drain outlet and moving upstream. For the chosen alignment, we note the locations at which natural subdrains enter into the main drain. These sections between themselves define specific reaches along the main drain. We can then estimate the catchment areas reachwise, in order to further estimate the peak discharge contributions received from the subdrains discharging into the main drain. These peak rates of flow would depend on the rainfall intensities, soil composition and land use pattern of the subcatchments. The bed level and the maximum flood water level along the open drains are to be properly fixed. From the point of ease of excavation, the bottom level should preferably be above the ground water table (if subsurface drainage is not an objective). Also the maximum water level should be low enough as to provide receptive outlets to the feeding subdrains. For this purpose the longitudinal bed slope, bed widths, etc, are suitably chosen to limit the maximum water level when the drain flows full.

The resulting water surface slope can be steeper than in the case of irrigation canals. Whereas the assurance of non-silting velocity is the major criterion in the design of irrigation canals, non-weeding velocity which sets the lower limit on velocity in order to inhibit the growth of weeds is also to be considered. The non-weeding velocity value can be 20 to 70% higher than the non-silting velocity. Usually the bed slopes in upstream reaches of the drain tend to be steeper and as we move down, the bed slopes adopted can be milder. The slope usually ranges between 1 in 1500 and 1 in 3000.

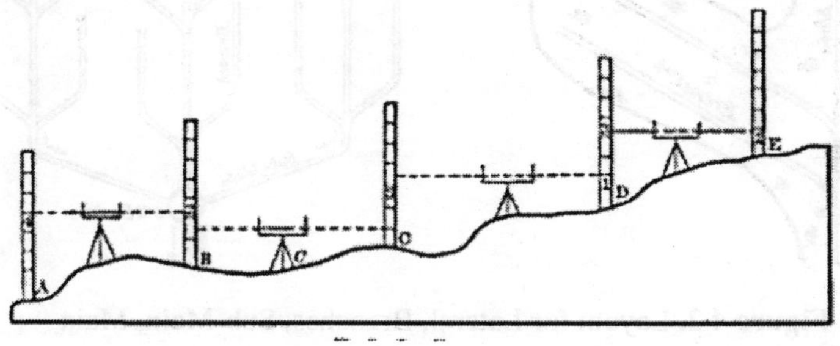

Figure 4.1. Levelling

4.5. LEVELLING FOR THE DRAINS

Mains

Having found the difference of elevation of various portions of the land to be drained, we are prepared to fix upon the lines for the main drains. At such times water will be found standing above the surface in hollows or basins in the land, and also on flats which seem as high as the surrounding surface. Mark these places and determine, if possible, whether the water is held by a clay sub-soil, or by the quantity of water retained in the soil at lower portions of the field. In the first case the natural drainage will be very slow, even though the elevation be sufficient; while in the latter, the natural drainage will go on rapidly if the surplus water is removed from the lower portions of the field, thereby giving the water an outlet through the soil.

If the whole field seems nearly flat, see if there are not some spots which are wetter than others, though the contour of the surface does not indicate it. Upon examination it may be found that the cause of this is with the sub-soil, as before noticed, or with the soil itself, it being made up largely of clay and more retentive of water.

The general rule for the location of mains is to let them follow the lowest land, or course of natural drainage. The surface then slopes towards the drain, making both natural and artificial drainage easy. We might say here that in all cases we should try and take every advantage that nature has given us in this work, for artificial drainage is only completing the work which nature has begun.

There are cases which require us to make exceptions to the general rule just given. First, the drain should be as free from angles and short turns as possible. In other words, it should be laid on a straight line, or a series of straight lines, connected by long curves. A few words in explanation of the advantage of straight lines and easy curves will convince the reader of their importance.

Short curves decrease the velocity of flow, so that, if we wish to have a uniform velocity in all parts of the drain, the grade must be increased at the bend, and as a consequence, the grade of the entire line must be lessened.

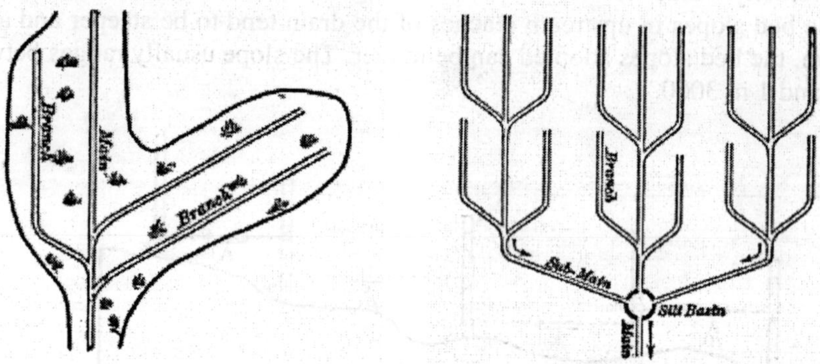

Figure 4.2. Layout for Lateral, Branches, Sub-Main, Main

Sub-mains and Branches

It is often the case that a single line of tiles laid through a flat, basin or hollow will afford a sufficient drainage for the purpose of the farmer. It consists of parallel branch drains discharging into the main, often at right angles to it. The junction of all branch drains with mains and sub-mains should be at such an angle that the pipe will discharge as nearly as possible in the direction of the current of the main and larger stream. Where it is necessary to have the drains connect at right or obtuse angles, the junction tiles should be curved in the direction of the current into which it discharges.

Begin at the outlet of the main and drive a grade peg about one foot to the right of the center line of the drain, and as near the end as is desirable. At the right of the peg and about live inches from it, set a guide stake, near the top of which mark O with a heavy lead pencil. Measure a length of fifty feet from the O stake and set a grade peg and accompanying guide stake, as in the first case, and mark the figure 50 upon the guide stake. The upper marks upon the guides indicate the number of feet of length from the outlet. The grade pegs should be placed as near to the edge of the proposed ditch as they can and remain firm after the ditch is dug. There is no reason why the stakes should be set fifty feet apart rather than any other distance, except that in practice it seems to be the most convenient to work from. Let the whole main line be staked out in this manner, numbering the stakes 0, 50, 100, 150, etc. Where there is a curve in the ditch let the chain or tape follow the line, that we may know the exact number of tiles required.

Another thing to be noticed at the time the main is staked out is, where the sub-mains and branches enter. Wherever a junction is made a grade peg and guide should be placed. This peg marks the outlet of another drain, and should be marked O, and its name in addition to its number, on the main line.

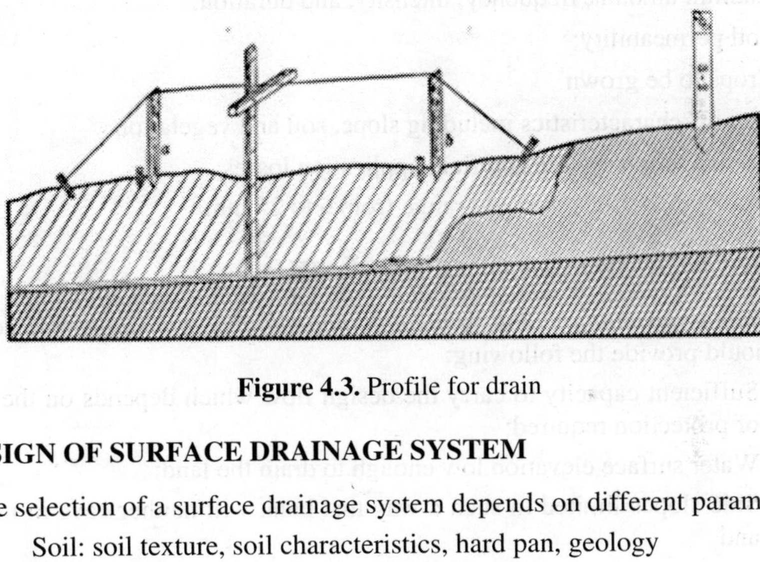

Figure 4.3. Profile for drain

4.6 DESIGN OF SURFACE DRAINAGE SYSTEM

The selection of a surface drainage system depends on different parameters like:

1. Soil: soil texture, soil characteristics, hard pan, geology
2. Topography
3. Climate: Rainfall intensity, frequency, etc.
4. Crop factors: Types of crop, root system, etc.
5. Cost of land and
6. Availability of outlet.

The land to be drained should be capable of providing good permanent cropland or pasture after drainage. The economic justification of drainage will depend largely upon the value of the crops that can be grown on the drained land. The three main types of surface drainage are open ditches, grassed channels and land grading. Shallow open drainage ditches are often used in conjunction with land grading and leveling to provide a complete practical solution to a drainage problem. Shallow open ditches can be used to remove surface water or intercept the overland flow, while deep open ditches can be used to provide subsurface drainage. The spacing and depth required for deep open ditches is similar to subsurface drains. Because of this, subsurface drains are usually the most feasible solution to lower watertable and are often used in conjunction with surface ditches.

The rate of drainage expressed as the depth in centimetres of water drained off from a given area in 24 hours, is called the drainage coefficient or drainage design rate. It is the design value at which water is to be removed from a drainage area. It may also be expressed in terms of flow rate per unit area, as cubic metres per square kilometre per 24 hours, or in terms of the flow rate per unit of area, which varies with the size of the area.

The drainage coefficient depends on:

- Geographical location
- Land use
- Size of area
- Rainfall amount, frequency, intensity, and duration;
- Soil permeability;
- Crops to be grown
- Run off characteristics including slope, soil and vegetation
- The irrigation system and field application losses,

The drainage coefficient would range from 6 to 25 mm.

4.7. HYDROLOGIC DESIGN OF OPEN DRAIN

Drainage requirements: An open ditch properly designed to provide adequate drainage should provide the following:

(1) Sufficient capacity to carry the design flow which depends on the degree of protection required;

(2) Water surface elevation low enough to drain the land;

(3) Side slopes limited in such a way that bank will neither cave in nor slide and

(4) Velocity of flow such that neither serious scouring (removal) nor silting (deposition) of any considerable extent will occur.

Estimating the Design Flow Rate

The first step in surface drainage system design is to calculate the watershed area and the peak runoff rate. The design flow rate is the excess water to be drained and the rate at which it has to be drained to avoid the waterlogging effect on crops and on the soil. The design involves calculating the drainage channel geometry and the surface drainage network layout. Considerations of the existing outlet condition, the condition of the disposal system and selection of suitable outlet.

There are two stages in the design of surface drainage: (1) the hydrologic design for estimating or establishing the design drainage coefficient and (2) the hydraulic design for facilitating the drainage. The objective of peak runoff estimation is to provide necessary data for the design of agricultural surface drainage systems. It should focus on the amount of water that can be expected and its frequency. The determination of the capacity of drainage channels for the safe disposal of excess water from the surface of the land is an important and at the same time difficult task in planning a drainage scheme. The volume of water that must be removed by surface drainage channels is dependent upon a number of variable conditions like Intensity of rainfall; Soil characteristics in the watershed; Size and surface topography of the watershed area; Vegetative cover and land use in the watershed; and Tolerance of crops to flooding in the area to be protected.

An open ditch properly designed to provide adequate drainage should provide the following: (a) Sufficient capacity to carry the design flow which depend on the degree of protection required; (b) Water surface elevation low enough to drain the land; (c) Side slopes limited in such a way that banks will neither cave in nor slide; and (d) Velocity of flow such that neither serious scoring (removal) nor silting to any considerable extent will occur.

When water entering an area is more than what can be transmitted or routed by established watercourses, flooding occurs. Generating hydrographs from big storms becomes the job of flood prediction. We use the word *prediction* because we still deal with probabilities of events.

Some of the reasons we study runoff peaks and volumes are:

- To quantify the volume and rate of water to be handled by water management facilities
- To predict soil erosion and transport of surface pollutants
- To identify critical non-point source pollution areas

Peak runoff rate

Definition: The maximum volume flow rate passing a particular location during a storm event. Peak discharge has units of volume/time (e.g. ft^3/sec, m^3/sec, acre-feet/hour).

The peak discharge is a primary design variable for the design of stormwater runoff facilities such as pipe systems, storm inlets and culverts, and small open channels. It is also used for some hydrologic planning such as small detention facilities in urban areas.

There have been many different approaches for determining the peak runoff from an area. As a result many different models *(equations)* for peak discharge estimation have been developed.

Ideally, we would like to have a 50-year flood record available at every site where a peak discharge estimate is needed for design work. If such data were always available, then a frequency analysis of the flood record could be used to characterize the flood potential at the site of the design work.

More often than not, flood records are rarely available where peak discharge estimates are needed for design work. Therefore, it is necessary to use either a prediction method that was developed from flood frequency analyses of gauged data in the region or an uncalibrated prediction equation that was designed for use at ungauged sites.

4.8. RATIONAL METHOD OF PREDICTING PEAK RUNOFF RATE

The most widely used uncalibrated equation is the Rational Method. Mathematically, the rational method relates the peak discharge (q, m^3/sec) to the drainage area (A, ha), the rainfall intensity (i, mm/hr), and the runoff coefficient (C).

SI Units $q = 0.0028CiA$

Where

q = design peak runoff rate in m^3/s

C = the runoff coefficient

i = rainfall intensity in mm/h for the design return period and for a duration equal to the "time of concentration" of the watershed.

English Units $q = CiA$

Where q – ft^3/sec

i = rainfall intensity (in/hr)

A = watershed area in acres

C = runoff coefficient, ratio of the peak runoff rate to the rainfall intensity, dimensionless

To use the rational method there are a few assumptions.

- Rainfall intensity and duration is uniform over the area of study
- Storm duration must be equal to or greater than the time of concentration of the watershed.

Time of Concentration (T_c)

Definition – The time required for water to flow from the most remote (in time of flow) point of the area to the outlet once the soil has become saturated and minor depressions filled. It is assumed that when the duration of the storm equals the time of concentration, all parts of the watershed are contributing simultaneously to the discharge at the outlet. There are several methods for computing time of concentration. Two of the most popular methods are the Kirpich equation and the SCS lag formula.

Kirpich Formula

$$T_c = 0.0195\, L^{0.77} S^{-0.385}$$

where

T_c = time of concentration in minutes.

L = maximum length of flow in meters.

S = the watershed gradient in m per m or the difference in elevation between the outlet and the most remote point divided by the length L.

Note – to convert to English units, the constant becomes 0.0078.

SCS Lag Formula

$$T_c = 0.00526\,L^{0.8}(1000/CN - 9)^{0.7}\,S^{-0.5}$$

where, L = watershed length (ft)
S = watershed slope (ft/ft)
CN = curve number

Rational Method Runoff Coefficients

The rational method uses runoff coefficients in the same fashion as the SCS curve number method for estimating runoff volume. They have been determined over the years and primarily focus on urban watershed applications. Below are several tables for different land conditions.

Table 4.2. Runoff Coefficient "C" for Agricultural Watersheds (Soil Group B)

Cover and hydrologic condition	Coefficient C for rainfall rates of		
	25 mm/h (1 iph)	100 mm/h (4 iph)	200 mm/h (8 iph)
Row crop, poor practice	0.63	0.65	0.66
Row crop, good practice	0.47	0.56	0.62
Small grain, poor practice	0.38	0.38	0.38
Small grain, good practice	0.18	0.21	0.22
Meadow, rotation, good	0.29	0.36	0.39
Pasture, permanent, good	0.02	0.17	0.23
Woodland, mature, good	0.02	0.10	0.15

Table 4.3. Hydrologic Soil Group Conversion Factors

Cover and hydrologic condition	Factors for converting the runoff coefficient C from group B soils to"		
	Group A	Group C	Group D
Row crop, poor practice	0.89	1.09	1.12
Row crop, good practice	0.86	1.09	1.14
Small grain, poor practice	0.86	1.11	1.16
Small grain, good practice	0.84	1.11	1.16
Meadow, rotation, good	0.81	1.13	1.18
Pasture, permanent, good	0.64	1.21	1.31
Woodland, mature, good	0.45	1.27	1.40

"Factors were computed by dividing the curve number for the desired soil group by the curve number for group B

Table 4.4. Runoff Coefficients for the Rational Formula versus Hydrologic Soil Group (A, B, C, D) and Slope Range

Land Use	Hydrologic Soil Group (A, B, C, D) and Slope Range											
	A			B			C			D		
	0-2%	2-6%	6%+	0-2%	2-6%	6%+	0-2%	2-6%	6#+	0-2%	2-6%	6%*
Cultivated land	0.08'	0.13	0.16	0.11	0.15	0.21	0.14	0.19	0,26	0.18	0.23	0.31
	0.14"	0.18	0.22	0.16	0.21	0.28	0.20	0.25	0.34	0.24	0.29	0.41
Pasture	0.12	0.20	0.30	0.18	0.28	0.37	0.24	0.34	0.44	0.30	0.40	0.50
	0.15	0.25	0.37	0.23	0.34	0.45	0.30	0.42	0.52	0.37	0.50	0.62
Meadow	0.10	0.16	0.25	0.14	0.22	0.30	0.20	0.28	0.36	0.24	0.30	0.40
	0.14	0.22	0.30	0.20	0.28	0.37	0.26	0.35	0.44	0.30	0.40	0.50
Forest	0.05	0.08	0.11	0.08	0.11	0.14	0.10	0.13	0.16	0.12	0.16	0.20
	0.08	0.11	0.14	0.10	0.14	0.18	0.12	0.16	0.20	0.15	0.20	0.25
Residential lot	0.25	0.28	0.31	0.27	0.30	0.35	0.30	0.33	0.38	0.33	0.36	0.42
size 1/8 acre	0.33	0.37	0.40	0.35	0.39	0.44	0.38	0.42	0.49	0.41	0.45	0.54
Residential lot	0.22	0.26	0.29	0.24	0.29	0.33	0.27	0.31	0.36	0.30	0.34	0.53
size 1/4 acre	0.30	0.34	0.37	0.33	0.37	0.42	0.36	0.40	0.47	0.38	0.42	0.52
Residential lot	0.19	0.23	0.26	0.22	0.26	0.30	0.25	0.29	0.34	0.28	0.32	0.39
size 1/3 acre	0.28	0.32	0.35	0.30	0.35	0.39	0.33	0.38	0.45	0.36	0.40	0.50
Residential lot	0.16	0.20	0.24	0.19	0.23	0.28	0.22	0.27	0.32	0.26	0.30	0.37
size 1/2 acre	0.25	0.29	0.32	0.28	0.32	0.36	0.31	0.35	0.42	0.34	0.38	0.48
Residential lot	0.14	0.19	0.22	0.17	0.21	0.26	0.20	0.25	0.31	0.24	0.29	0.35
size 1 acre	0.22	0.26	0.29	0.24	0.28	0.34	0.28	0.32	0.40	0.31	0.35	0.46
Industrial	0.67	0.68	0.68	0.68	0.68	0.69	0.68	0.69	0.69	0.69	0.69	0.70
	0.85	0.85	0.86	0.85	0.86	0.86	0.86	0.86	0.87	0.86	0.86	0.88
Commercial	0.71	0.71	0.72	0.71	0.72	0.72	0.72	0.72	0.72	0.72	0.72	0.72
	0.88	0.88	0.89	0.89	0.89	0.89	0.89	0.89	0.90	0.89	0.89	0.90
Streets	0.70	0.71	0.72	0.71	0.72	0.74	0.72	0.73	0.76	0.73	0.75	0.78
	0.76	0.77	0.79	0.80	0.82	0.84	0.84	0.85	0.89	0.89	0.91	0.95
Open space	0.05	0.10	0.14	0.08	0.13	0.19	0.12	0.17	0.24	0.16	0.21	0.28
	0.11	0.16	0.20	0.14	0.19	0.26	0.18	0.23	0.32	0.22	0.27	0.39
Parking	0.85	0.86	0.87	0.85	0.86	0.87	0.85	0.86	0.87	0.85	0.86	0.87
	0.95	0.96	0.97	0.95	0.96	0.97	0.95	0.96	0.97	0.95	0.96	0.97

* Runoff coefficients for storm recurrence intervals less than 25 years

The general procedure for determining peak discharge using the Rational Method is as follows:

Step 1: Determine the drainage area (in hectares).

Step 2: Determine the runoff coefficient (C).

Step 3: Determine the hydraulic length or flow path that will be used to determine the time of concentration.

Step 4: Determine the time of concentration (T_c) for the drainage area.

Step 5: Determine the intensity using the time of concentration. The intensity is determined using the time of concentration and the I-D-F curve for the county.

Step 6: Input the drainage area, C value, and intensity into the formula to determine the peak rate of runoff.

The Rational method is used to predict design flows from small urban and rural watersheds. For rural watershed areas, the upper limit for which the Rational method is valid is 500ha. Two significant limitations to the Rational method are estimating the time of concentration and selecting an appropriate runoff coefficient. The assumptions of rational method are: the rainfall intensity used should be uniform over the entire watershed; the peak runoff rate occurs when the entire watershed is contributing.

4.9. ESTIMATION OF RUNOFF USING CURVE NUMBER TECHNIQUE

Runoff generated from storm rainfall is largely determined by local interaction of the properties of the rainfall, ground cover, land use, and soil. The predominance of relatively short high intensity storms in most sub humid and semi-arid zones means that these areas are more prone to infiltration excess runoff (Smith and Goodrich, 2005).

The US Soil Conservation Service developed the Curve Number method for application to large agricultural areas to predict surface runoff from daily rainfall data. The CN, an integer varying in between 0 and 100 of a geographical area is an integrated indicator of hydrological soil type, land use and land cover, ground surface conditions and Antecedent Moisture Condition (AMC). This method is used to estimate direct runoff volume from the rainfall depths. Antecedent Moisture Condition gives daily moisture or wetness condition of the soil before occurrence of storm. The volume of runoff depends on the volume of precipitation and the volume of storage that is available for retention.

Estimation of Curve Number (CN)

The runoff coefficient used for estimating surface runoff depth from rainfall depth. The CN incorporates the effects of infiltration characteristics of the soil, land use and agricultural practices. Appendix - C represents various values of CN for various soil groups and land use patterns for AMC - II. From this table it can be estimated for the different hydrologic conditions of the soil, the curve numbers are different. So there is need to estimate wetted curve number of the different soil. To estimate the wetted curve number equation is used.

$$CN = \frac{(A_1 \times CN_1) + (A \times CN_2)}{A_1 + A_2}$$

Where, A_1 and A_2 are the areas of different fields and CN_1, CN_2 are the curve number of respective fields.

In the SCS method, the change in potential maximum retention (S) is based in the index of the watershed wetness on the day of the storm, known as antecedent moisture condition, determine by the total rainfall in the five day period preceding a storm. There is a multiplying factor to convert AMC II condition to the AMC - I and AMC - III as shown in Appendix - A. There levels of AMC are used as,

AMC – I This level presents the lower limit of moisture or the upper limit of S. The soil is dry enough for satisfactory ploughing or cultivation.

AMC – II This level presents the average condition.

AMC – III This level presents the upper limit of moisture or the lower limit of S. The soil is practically saturated from antecedent rains.

Calculation of the potential maximum retention (S)

The formula used for the calculation of potential maximum retention (S) is as in equation.

$$CN = \frac{2540}{25.4 + S}$$

Where,

S = Potential maximum retention of rainwater at any time in cm

CN = Runoff Curve Number which representation the combined effect of soil, land use, hydrologic condition and antecedent soil moisture.

Calculation of initial abstraction (I_a)

Initial abstraction (I_a) which consists of interception losses, depression storage and infiltration before the runoff begins. Relationship between I_a and S was developed by means of rainfall and runoff data. The empirical relationship is given by

$$I_a = 0.1S \quad \text{for AMC II and AMC III}$$

$$I_a = 0.3S \quad \text{for AMC I}$$

If the amount of precipitation is less than or equal to 0.3 times of S in case of AMC I or 0.1 times of S in case of AMC II and AMC III, than there will be no runoff. Therefore, runoff is calculated for a rainfall greater than 0.3 S (AMC I) or 0.1 S (AMC II and AMC III).

Based on the catchment's water retention behavior, two parameters are used, namely potential abstraction and initial abstraction, I_a is taken as fraction of S namely, I_a = 0.1*S, 0.2*S or 0.3*S. The central Unit for Soil Conservation, Ministry of Agriculture, Government of India has suggested a modification in the I_a as follows:

- In India it is assumed as I_a = 0.1*S for black soil regions under AMC*II and AMCIII.*
- For black soil regions under AMC *I and* for all other regions, I_a = 0.3*S.

Calculation of surface runoff depth

The values of S and I_a have been put into the equation to get the runoff value on daily basis, which calculates the runoff in cm being given precipitation in cm.

$$Q = \frac{(P-0.2S)^2}{(P+0.8S)}, \quad Q = \frac{(P-0.1S)^2}{P+0.9S}, \quad Q = \frac{(P-0.3S)^2}{P+0.7S}$$

Where, Q = Runoff depth, cm

P = Rainfall depth, cm

S = Potential maximum retention of rainwater at any time, cm

I_a = Initial abstraction

The Curve Number (CN) method also known as the Hydraulic Soil Cover Complex Method widely used for runoff estimation. Several properties of the watershed are taken into consideration and their runoff producing capability is expressed by a numerical value varying between 0 and 100 then by using simple empirical formulae runoff is estimated for a given value of rainfall.

- Design rainfall
- In agricultural areas, the time of concentration is less than 6 hours.
- The standard practice is to take a 6-hr rainfall of a given or desired recurrence interval, which is normally taken as 5 yrs or 10 yrs.
- The 24-hr rain of the desired recurrence interval is converted into 6-hr rainfall by multiplying by 0.6 (0.5 to 0.7). The resulting value is distributed within 6 hours as follows, using the World Meteorological Organization guidelines.

Table 4.5. Distribution of 6-hr. rain within 6 hours

Time	% of rain	Time, hr	% of rain
0	0	3.5	78
0.5	2	4.0	84
1	8	4.5	8
1.5	15	5.0	92
2	22	5.5	96
2.5	60	6.0	100
3	70	-	-

Table 4.6. Hydrologic Soil Cover Complex

Group	Infiltration rate	Occurring in	Runoff potential
A	High(> 25 mm/hr)	Deep soil	Low
B	Moderate(12.5-25 mm/hr)	Deep coarse soils	Moderately low
C	Low(2.5-12.5 mm/hr)	Shallow fine soil	Moderately high
D	Very low(<2.5 mm/hr)	Clayey shallow soil	High

Antecedent Moisture Content (AMC): Based on preceding 5-day rainfall total, the catchment may be grouped under AMC *I, AMC II, AMCIII.*

Table 4.7. Antecedent Moisture Content (AMC)

AMC group	Preceding 5-day rain	Season	Drainage area	Expected runoff
I	<12 mm	dormant	Dry	Less
	<36 mm	growing		
II	12-38 mm	dormant	Relatively wet	Medium
	36-53 mm	growing		
III	>28 mm	dormant	Very wet	Heavy
	>53 mm	growing		

Hydrologic condition: Based on land coverage, the hydrologic condition of an area is said to be as under.

Table 4.8. Hydrologic condition

Hydrologic condition	Area covered by canopy	Runoff
Poor	<50%	Heavy
Fair	50-75 %	Medium
Good	>75%	Less

Table 4.9. Curve Number Values for Different Land Use and Cover Conditions

Land use	Cover condition	Antecedent soil moisture condition (AMC*II*)			
		$I_a = 0.3*S$		$I_a = 0.1*S$	
		Hydrologic soil group			
		A	B	C	D
Fallow	Poor	77	86	91	94
Row crop straight	Poor	72	81	88	91
	Good	67	78	85	89
contour	Poor	70	79	81	88
	Good	65	75	82	86
Terraced	Poor	66	74	80	82
	Good	62	71	78	81
Forest	Dense	26	40	58	61
	Open	28	44	60	64
Pasture	Poor	68	79	86	89
	Fair	49	69	79	84
	Good	39	61	74	80
Orchard	Average	40	54	68	72
Paddy	-	95	97	95	95

Table 4.10. Conversion of curve numbers from AMC-II to AMC-I and AMC-III

AMC	Curve Number values										
I	100	78	63	51	40	31	22	15	9	4	0
II	100	90	80	70	60	50	40	30	20	10	0
III	100	96	91	85	78	70	60	50	37	22	0

4.10. HYDRAULIC DESIGN OF OPEN DRAIN

Design Carrying Capacity

The return period of design rainfall could be chosen to range from 10 to 15 years. The peak discharge contributions entering various reaches of the drainage system can be estimated using the concept of drainage coefficient.

Drainage coefficient is the minimum rate at which the runoff water is to be led away by a drain so that the period of inundation of a cultivated area is short enough as not to affect the yield of crops. The tolerable duration of submergence varies between 3 to 5 days for most crops. Drainage coefficient is expressed as depth of water to be removed in 24 hours. Its value depends on rainfall frequency, its duration and intensity, type of crop, hydraulic conductivity of the saturated soil as discussed the course entitled Hydrology. In Punjab and Haryana, the average value of drainage capacity is 4 cumec per square kilometer. In Uttar Pradesh, it is low at 0.11 cumec/sq km. In Andhra Pradesh and Tamil Nadu, the drainage coefficient is considered to vary with the area drained; consequently a 10 sq km area would generate 4.4 cumec, a 25 sq km area would generate about 10 cumec, and a 50 sq km, are about 16 cumec. The maximum tolerable period of inundation in deltaic area of Tamil Nadu is taken as 7 days.

The design peak runoff rate of the drain is predicted after using the Rational formula. The water surface in the ditch should be kept about 30 cm below the surface of the fields to be drained.

Design Discharge

The dimensions of drains, whether they be open drains or pipe drains, are based on the required design discharge. This design discharge is influenced by the storage capacity of the drainage system. By reducing ponding or waterlogging, a drainage system creates a buffer capacity in the soil, ensuring that the discharge is steadier and smaller than the recharge.

If the soil has a large buffer capacity, a longer period of critical duration can be adopted and average recharge and discharge rates over this longer period can be used. In contrast, if the soil has only a small buffer capacity, the infrequent, extreme, recharge and discharge rates have to be assessed and shorter periods of critical duration have to be adopted.

Subsurface drainage systems create a medium storage capacity. In regions with low rainfall intensities (say less than 100 mm/month) and in irrigated lands in arid or semi-arid regions, the design discharge for the month or season with the highest net recharge has to be calculated.

In regions that have seasons with high rainfall (say more than 100 mm per month), it is likely that the problem is one of surface drainage rather than of subsurface drainage. Here, a subsurface system would not be appropriate, or it could be combined with a surface system. In a combined system, the design discharge of the subsurface system has to be calculated from a water balance after the discharge from the surface system has been deducted.

A surface field drainage system, consisting of beds in flat lands or mildly graded field slopes in undulating lands, creates only small capacities for storage. The design discharge must then be based on the water balance over a short period (say 2 to 5 days).

Slopes of Field Drains

The maximum slope of field drains is dictated by the maximum permissible flow velocity. If the topography should call for steep slopes, drop structures should be built into the drains. For pipe drains, these are normally incorporated in manholes.

Special caution is needed if a steep slope changes to a flatter slope: high pressures may develop at the transition point unless the flow velocity on the upstream side is properly controlled and the downstream (flatter) reach has a sufficient capacity to flow would occur in case caving occurs in such narrow channels. Wherever possible, it is desirable to check with the local practice, and adopt the side slopes prevailing in the channels existing in the area.

Ditch cross-section.

It is usually designed with trapezoidal cross sections.

Design Velocity

The value of design velocity to be maintained while draining water varies with the nature of soil in which the drain flows: basically, it has to be non-silting as well as non-scouring, and if possible should inhibit weed growth. In sandy soil, it would be the least at 0.75 m/s; for silty soil it is about 0.9 m/s, and for stiff clayey soil it is about 1.5 m/s. For course gravely soil, it could be as high as 2.5 m/s. Especially at low flows, there should be sufficient velocity which would inhibit sediment deposition and growth of aquatic plants. As a general guidance, velocities above 0.3 m/s would prevent silt and fine sand deposits, 0.6 m/s and above would prevent weeds and grasses to grow and velocities above 0.75 m/s would inhibit growth of aquatic plants. Manning's formula is commonly used relating velocity with other parameters, such as:

$$v = \frac{1}{n} R^{\left(\frac{2}{3}\right)} S^{\left(\frac{1}{2}\right)}$$

Where, v = Velocity of flowing water, m/sec

n = Manning's roughness coefficient

R = Hydraulic mean radius, m

S = Slope of the bed, m/m

Table 4.11. Value of Manning's n

Sr. No.	Bottom width	Manning's n
1	<1.2 m	0.040
2	>1.2 m	0.035

The permissible velocities in open drains for different soil textural groups are given in following table

Table 4.12. The Permissible Velocities in Open Drains

Sr. No.	Soil type	Maximum velocity, m/sec
1	Sand and sandy loam	0.75
2	Silt loam	0.90
3	Sandy clay loam	1.00
4	Clay loam	1.20
5	Stiff clay	1.50

A particular form of Chezy's equation (called Elliot's formula) may also be used. It is expressed as:

$$V = 0.158\sqrt{\frac{A}{P}(\Delta h)}$$

in which, V is the design velocity in m/s, A is the sectional area (sqm) of the drain, P is the wetted perimeter (m) and Δh is the difference in bed elevation over a kilometer length, expressed in centimeter. This formula assumes a value for Chezy's C as 50 $[m^{1/2}/S]$.

Since Manning's $n = \frac{1}{C} R^{1/6}$, the corresponding value of n would vary in the range of 0.02 to 0.028, for the range of value of the hydraulic radius, R = 1 to 8 m.

Side slope

Depending on the type of soil layer through which the drain is to be cut, the side slopes may vary from almost vertical to about 3H: 1V.

Table 4.13. Recommended Side Slopes for Drainage Channel

Sr. No.	Soil type	Slope
1	Clay	1:1
2	Silt loam	1 ½ :1
3	Sandy loam	2:1
4	Loose sandy loam	3:1
5	Hard pan	½ :1
6	Peat and muck	¼ : 1

Channel side slopes, in fact, are seen to depend on soil texture with a view to stability of the sides of a drain. The most critical condition arises when the soil of the side slopes is left saturated soon after a rapid drop in the level of water, and the sides then cave in (You may recall that in the case of earth dams also, one critical condition is sudden draw-down). For the same side slopes, a deeper ditch is more likely to cave in. In the case of narrow ditches (for which the bed width is smaller than 0.75 m, say) flatter side slopes than indicated in Table are preferable since serious obstruction.

Most Efficient Bottom Width

For the most efficient cross-section and minimum volume of excavation, the bottom width is determined by the following formula.

$$b = 2 \times d \times \tan(\theta/2)$$

Where;

b = bottom width;

d = depth;

θ = side slope angle.

Exercises

1. Find out the value of CN and estimate the drainage coefficient of a catchment 200 ha of flat agricultural land having 100 ha of row crops in good condition, laid out on contour, 20 ha under pasture, 30 ha under forest, 30 ha under orchard and 20 ha under paddy. The forest is denuded and the pasture is overgrazed. The 5-yr 1- day rainfall recorded is 140 mm. The preceding 5-day rain is 60 mm. The hydrologic group is C.
2. Design the most efficient bottom width for a drainage channel to carry a flow 2 m deep in the clay soil. Compute the velocity and discharge capacity of the channel if the channel gradient is 0.04%.
3. Find out the dimension of a drainage canal to drain 550 ha of land having a drainage coefficient of 2.5 cm. The soil is silt loam. Maximum permissible slope of channel bed is 0.1%.
4. Design the main open drain with bed slope of 0.1%, assuming the value of the Manning's n to be 0.03. The drainage coefficient of an area is 66 mm/day. The catchment area is 6 km^2. The type of soil will permit a side slope of 1.5:1. The depth of the drain at the outlet cannot exceed 1.5 m, based on the topography of the area and the elevation of the outlet.
5. Estimate the rate of design discharge in cumec for a surface drain in Gujarat that has a drainage area of 2500 ha upto a certain chainage, if the corresponding empirical formula is q = (4.8 + 0.2 A), wherein q is in m3/s and, 4 is in sq km.
6. The cross-section of a drainage channel has a bed width of 1.75 m, a side slope of 1 : 1 and a depth of flow of 2 m. For a drainage coefficient of 12.5 mm/day/ha, compute the area that can be safely drained by this section, at a bed slope of 1 in 4000, as per Mannings equation with n = 0.022.

CHAPTER - 5

SUBSURFACE DRAINAGE SYSTEM AND DESIGN

A subsurface drainage system induces a system of pipes or open drains to remove excess water and dissolved salts. Sub-surface drains are underground artificial channels through which excess water may flow to a suitable outlet. Their purpose is to lower the ground water level below the root zone of the plants by taking out only the excess water through action of gravity. Drainage requirements for optimum plant growth are determined largely by the volume and content of soil air which get reduced in presence of excess water. These requirements depend on type of crop, soil, availability of plant nutrients, climatic conditions, biological activities and soil and crop management practices. The rooting depth and the tolerance to excess water are the most important crop characteristics. For example, you are aware of the fact that rice is tolerant because it has special internal mechanisms for obtaining oxygen from the air, whereas tobacco is sensitive to excess water. So based on crop need, priorities of drainage programmes are fixed. Apart from providing good soil aeration, drainage has many other benefits e.g. removal of toxic substances or allowing timely cultural practices of crop production. However, the benefits of drainage can be realised only when the soil is potentially productive, if drained. In northern India, which enshrined most of the productive regions of the country, water-logging and related adverse effects are compounding. Properly designed sub-surface drainage programmes keeping the cost-benefit ratio in backdrop, is a much needed activity.

5.1. BENEFITS OF SUB-SURFACE DRAINAGE

(a) Soil aeration

Soil air helps in maximum development of plant roots and desirable soil organism. Aeration of soil is the most important factor and plants are not adversely affected even in total water ponding if air is provided. In an undrained situation, a high water table does not allow roots to grow in early age of the plant. A better drained field provides better roots growth which, results in increased volume of soil from which root can obtain food. It also makes available more water to plant through deepened roots.

(b) Increased crop yield and production

The better drainage gives an increased length of growing season because of earlier possible planting dates. Better drainage facility provides better plant growth, which increases productivity. Under field conditions, soil water contents changes with time during plant growth from wilting point to complete saturation. It is useful to characterise the effects of water stress on crop yields for both irrigation and drainage requirement. Adding productive land (by reclaiming the land) without extending farm boundaries, Permitting good soil management, Permitting timeliness in sowing and harvesting of crop, and allowing efficient utilisation of farm equipment and machinery to accrue advantages of farm mechanisation.

(c) Improve soil physical condition

As sub-surface drainage removes excess moisture, it is easy to get better soil physical condition and thus suitable soil tilth for plant growth in the root zone. If sub-surface moisture level is high, there is possibility that soil tilth may get disturbed through operation of tillage equipment.

(d) Movement of machinery

The sub surface drainage helps in improvement of soil water conditions in relation to operation of tillage, planting and harvesting machines.

(e) Improves soil structure

It increases the bacterial activity in the soil which improves the soil structure and makes the plant food more readily available.

(f) Reduces soil erosion

It reduces soil erosion, since a well drained soil has great storage capacity for water, resulting in less runoff.

(g) Removes toxic substances

Sub-surface drainage helps in removal of toxic substances, such as salts. These salts, in certain soils retard plant growth. Death of the plant may be caused by toxic substances moving up from the roots. Also, when aeration is reduced, iron and manganese may become high enough in concentration to be toxic to the roots. Decomposition of organic matter may produce hydrogen sulphide which is also toxic.

(h) Oxygen and carbon dioxide balance

Flooding of plant roots causes a rapid reduction in transpiration, reduced absorption of oxygen and other plant nutrients with a corresponding increase in the carbon dioxide content of the soil, a disturbance of microbiological activity and a reduced mass of soil from which nutrient can be obtained. Changes in the oxygen and carbon dioxide balance may affect growth of disease organism in the soil. And lastly, with prolonged flooding and reduced biological activity, soil structure may be destroyed, which in turn, reduces aeration.

(i) Increased nitrogen supply

It also help in increased nitrogen supply from soil to plants which in turn results in increased crop yield. At high water table conditions, crops can not obtain nitrogen from soil, however, due to proper drainage nitrogen applied in soil gives a good response.

(j) Preservation of wild life habitat

The major impact of drainage has been related to the preservation of wild life habitat and non-point pollution of drainage water for downstream users.

(k) Reduction of flooding

The hydrologic impact of sub-surface drainage is the reduced flooding. Studies have shown that peak runoff rates from small, flat plots were reduced 7 per cent by installing sub-surface drains.

5.2 METHODS OF SUBSURFACE DRAINAGE

Subsurface drainage aims at controlling the water table, a control that may be achieved by tubewell drainage, open drains or subsurface drains (pipe drains or mole drains).

In planning and designing a subsurface drainage system, decisions are to be taken on the layout and arrangement of drain lines, selection of a suitable outlet, proper depth and spacing of laterals, gradients of collectors and conductors, length and diameter drain pipes, material of the drains to lend adequate strength, and accessories such as inlets and outlet structures.

A tile drainage system consists of a drainage outlet, tile main, sub-main and laterals. The laterals remove free water from the soil; and the submains and mains carry drainage water to the drainage outlet.

The following factors affect the movement of sub surface water;

(a) Permeability of the soil horizon

(b) Availability and location of free water, presence and magnitude of artesian pressure or back pressure in drain

(c) Depth and spacing of drains

(d) Drain diameter

(e) Tile joint spacing

(f) Depth to impervious layer below the ground surface.

All these factors are important while taking up design of tile drainage system. Design of tile drainage system should be such that it meets the functional and locational requirements. The design of tile drains includes:

(1) The layout and arrangement of drain line,

(2) Selection of suitable outlet,

(3) Proper depth and spacing of laterals to collect excess water efficiently,

(4) Determination of length and size of drains,

(5) Selection of good quality material for different segments of the system,

(6) Design of inlet and outlet structures.

The layout and arrangement of drains has been explained to you already. The performance of a drainage system and its longevity depends upon proper design, installation and maintenance of different parts of drainage system including layout and accessories like inlet and outlet structures.

5.3 TILE DRAIN SYSTEM

The design of a tile drain system needs determination of (1) layout of the system; (2) depth and spacing of the drains; (3) size of the drains; (4) material of the tiles; (5) grade of the tile lines; (6) envelope materials and accessory structures; and (7) outlets. You shall study these designs in the next unit.

The different layout followed are: (1) random or natural; (2) herringbone, (3) gridiron, and (4) the interceptor drains. In an area it is possible to use one or a combination of the systems depending upon the site conditions. The suitability of these layout is briefly described below:

(a) The random or the natural system is used for draining isolated patches when the entire area does not need drainage.

(b) It is economical to use herringbone system in areas where a mainline or submain could be laid both sides. The land in this system is well drained.

(c) The gridiron system is similar to the herringbone system except that the laterals enter the main only from one side. The system could be more economical than the herringbone system as less number of junctions are used.

(d) The interceptor or the cut-off drain is placed at the upper edge of a wet area to intercept the sub-surface seepage. Open drains are also used for the purpose but require considerable land area.

5.4 DEEP OPEN DRAINS

Deep open drains use land area as well as need regular maintenance. The open drains are used both as interceptor and relief drains. The use of deep open drains as interceptor drains is more common as their length required is generally less than the relief drains and they can be used for intercepting both surface and sub-surface flows. They are similar to diversion drains used in soil conservation practices. The capacity of the interception drains, when it is intercepting only ground water flow can be estimated using Darcy's law.

5.5 DEPTH AND SPACING OF TILE DRAINS

The depth and spacing of sub-surface drains is a function of movement of water into drains. A definite relationship exists between depth and spacing for the soils of uniform permeability; the deeper the drains wider the spacing; hence less number of drains are required. The key aim of a tile drainage design system is to provide adequate root depth above the saturated zone.

The depth and spacing of tiles apply only to laterals and not to mains and submains. The depth of mains and submains is governed by the outlet conditions and the topography of the drainage area. The optimum tile depth for laterals is influenced by

(i) The hydraulic conductivity of the soil,

(ii) The rate of infiltration during irrigation,

(iii) Tile spacing,

(iv) Type of crop and its optimum water table depth,

(v) Crop responses to fluctuating water table, and

(vi) Depth of impermeable strata.

The depth of the tile drains should be such that midway between the drains, the water table should be at a satisfactory depth. Where an impermeable layer is encountered the tiles should be placed above such layer, except when the impermeable layer occurs at very shallow depth. If the tiles are located below the impermeable layer, back filling of the trenches should be done with permeable soil. The depth and spacing are determined using different requirements in humid and arid region. In humid areas, as during heavy rainfall water table rises to near the surfaces, hence, the rate of drop of the water table is important. In arid region under irrigation, the main consideration is to maintain minimum depth of the watertable for optimum crop growth.

As mentioned above, in general, the deeper the drains the wider the spacing and less number of drains are needed. On the other hand, deeper drains will result in greater installation costs and may cause outlet problems.

The following table gives the general depth and spacing that could be adopted under different conditions.

Table 5.1. Average Depth and Spacing of Tile Drains (adopted from Schwab *et al*; 1993)

Type of Soil	Hydraulic Class	Conductivity (m/day)	Spacing (m)	Depth (m)
Clay	Very slow	0.001	9-15	0.9-1.1
Clay loam	Slow	0.001-0.005	12-21	0.9-1.1
Average loam	Moderately slow	0.005-0.02	18-30	1.1-1.2
Fine sandy	Moderate	0.02-0.06	30-37	1.2-1.4
Sandy loam	Moderately rapid	0.06-0.13	30-60	1.2-1.5
Peat & Muck	Rapid	0.13-0.25	30-90	1.2-1.5
Irrigated	Variable	0.025-25	45-180	1.5-3.0

The subsurface drainage systems are designed based on either steady state approach or non-steady state approach. In steady state approach, the design rate of excess water removal is the drainage coefficient. The depth of the tile drains should be such that midway between the drains, the water table should be at a satisfactory depth. Based upon the hydraulics of flow of water in soils several formulas for determination of depth and spacing of tiles are proposed.

Pipe depth, defines as the distance from the surface to the bottom of the pipe, varies in different soils. Under no conditions should the amount of cover over the top of the pipe be less than 0.6 m. This minimum depth is necessary to protect the pipe from heavy surface loads and to prevent shifting of the pipe. In uniformly permeable mineral soils the depth of laterals usually varies from 0.8 to 2.5 m unless limited by the impermeable layer, one should design for the maximum depth, as this will permit a wide spacing. In deep organic soils after initial settlement has taken place, the minimum depth should not be less than 1.2 m.

In arid regions under irrigation the drainage design criteria are determined more by the minimum depth of the water table for optimum crop growth than by the rate of drop. Depths of 2 to 3 m are more common. Selection of salt-tolerant crops, careful application of water, and proper soil management may permit successful with shallower depths than indicated.

5.6 STEADY STATE CONDITION

The rate of excess water removal in the steady state condition estimated corresponding to a condition of equilibrium between the water entering the soil profile from rainfall or irrigation and the depth of water table. The steady state analysis is easier and is usually adopted in drainage design. It includes the theory of seepage of ponded water. It may, however, be mentioned that a drainage theory describes the flow of water in an idealized soil water system. The actual field problem is simplified to make it possible to get a mathematical solution.

The steady-state design is based on the assumption that the rate of recharge to the groundwater is uniform and steady and that it equals the discharge through the drainage system. Thus, the watertable remains at the same height as long as the recharge continues.

Figure shows cross-sections of a drainage system where the groundwater is under recharge from excess rainfall, excess irrigation, or upward seepage, the water-table is curved, its elevation being highest midway between the drains. Because of the symmetry of the system, we only have to consider one half of the figure. To describe the flow of groundwater to the drains, the following assumptions are to be made:

- Two-dimensional flow. This means that the flow is considered to be identical in any cross-section perpendicular to the drains; this is only true for infinitely long drains;
- Uniform distribution of the recharge;
- Homogeneous and isotropic soils. Thus any spatial variation in the hydraulic conductivity within a soil layer can be ignored.

A simple equation, called Ellipse equation, based on Darcy's law and Dupuit assumptions was derived by Colding in 1872, relating the depth and spacing of drains. An important modification for cases in which an impervious base is at a finite depth was proposed by Houghoudt and Schifgaarde in 1963. A graphical representation of the variation of the equivalent depth (0 to 4 m) with actual depth of the impervious layer (0 to 10 m) for different spacings (6 to 90 m) for a 125 mm pipe was presented by Schilfgaarde. For other drain diameters. Moody developed, in 1966, expressions for the equivalent depth.

Steady state formulae

With a drainage system of known or specified depth of drain, drain spacing, depth of the underlying barrier (impervious layer) and hydraulic conductivity, the configuration and level of water table between the drains will get established at some height and it will retain that shape as long as the recharge rate continues at the assumed constant value. Even though recharge due to rainfall and also due to irrigation application is usually intermittent, the assumption of steady recharge has been adopted as a satisfactory first approximation for quite a few cases.

In the United States of America, Donnan formula for drain spacing is generally used. It states that

$$L = \frac{4\ K\ (b^2 - a^2)}{i_p}$$

where

L = drain spacing (m)

K = hydraulic conductivity (m/day)

a = height of the drain above the barrier (m)

b = height of the highest point on the water table between two adjacent drains measured above the barrier (m), and

i_p is that part of infiltration (m^3/m /day) that enters the surface and finally percolates down to join the water table.

It is essential to adopt an appropriate value for ip. It is usually obtained as the depth of deep percolation, due to peak irrigation, divided by the number of days between successive irrigations.

The so called Ellipse equation is based on the following assumptions :

(i) The soil is homogeneous and isotropic,

(ii) The system is underlain by a horizontal impermeable layer,

(iii) Steady state condition prevails with the rate of recharge into the medium ip being in balance with the drain-discharge rate, q. (Thus, ipx L x 1m = q per meter length, wherein L is the spacing between adjacent drains),

(iv) The drainage is into an open ditch excavated down to reach the impervious layer and the ditch water level is steady, and

(v) The water table shape is symmetrical between adjacent drains.

The drainage equations are based on the Dupuit-Forchheimer assumptions. These allow us to reduce the two-dimensional flow to a one-dimensional flow by assuming parallel and horizontal stream lines. Such a flow pattern will occur as long as the impervious subsoil is close to the drain. If the impervious layer does not coincide with the bottom of the drain, the flow in the vicinity of the drains will be radial and the Dupuit-Forchneimer assumptions cannot be applied.

Moreover, in this regard the **Dupuit-Forcheimer assumptions** are also adopted as given below:

(i) All streamlines may be treated as horizontal, since the spacing between the drains (L) is usually very large compared with hm (the maximum height of the water table above the center of the spacing between the drains, which would occur midway between adjacent drains). This assumption permits the use of hydrostatic pressure distribution in the flow domain; and

(ii) The velocity along any of the streamlines cutting a vertical section is the same as given by Darcy's equation, in which the slope of free water surface at that section is considered as the hydraulic gradient for all points on that vertical section. This implies a single value for the velocity at all the points, and also the velocity is treated as horizontal in direction.

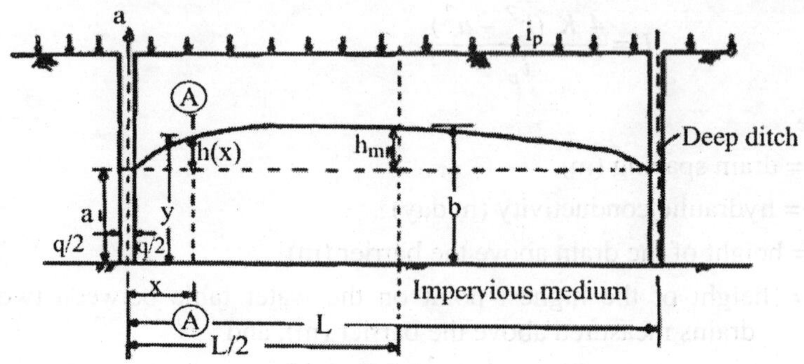

Figure 5.1. Steady Flow into a Ditch

Figure depicts a steady flow into a ditch considering a unit width normal to the section shown in the figure. We can write (as mentioned above):

$$q = (i_p \times L \times 1\,m)$$

$$q = (i_p \times L)$$

where,

q = the drainage rate per unit length of drain pipe,

ip = the recharge rate (in m^3/day/m^2).

It is clear that the length contributing to any one drain is made of a length of *L/2* on each side of the drain. Since i_p is uniform and steady, the seepage flow rate towards a drain should vary linearly from the mid point between two adjacent drains towards the drain. Thus the rate of drainage at any distance x from the drain edge represented as q_x can be expressed as:

$$q_x = \frac{\left(\frac{L}{2} - x\right)}{\frac{L}{2}} \times \frac{q}{2}$$

But then according to Darcy's law and Dupuit Forcheimer assumption, the seepage *(qx)* crossing a vertical section of unit width along the drain axis is given by:

$$q_x = \left(K\frac{dy}{dx}\right)(y \times 1)$$

Herein *K dy/dx* gives the velocity normal to the section as per Darcy's law, and the assumption as presented above allows us to use this single value for all points in the vertical section of the area (y x 1). It is to be noted that the seepage flow area varies with distance x. Equating the right hand side terms in equations, the differential equation for the free water surface, (that is, the water table) may be obtained, in variables separable form, as given below

$$y\,dy = \frac{q}{KL}\left(\frac{L}{2} - x\right)dx$$

Integrating between limits x = 0 to x = x (an arbitrary value) with corresponding y values as a (say), elevation with respect to an impervious layer and y, we have

$$\frac{1}{2}y^2 \Big|_a^y = \left(\frac{q}{KL}\right)\left(\frac{L}{2}x - \frac{x^2}{2}\right)\Big|_0^x$$

$$\left(y^2 - a^2\right) = \left(\frac{q}{KL}\right)\left(Lx - x^2\right)$$

Now, midway between the drains (x = L/2) we represent the value y = b. Therefore, midway between the drains we have

$$\left(b^2 - a^2\right) = \left(\frac{q}{KL}\right)\left(\frac{L^2}{4}\right)$$

Replacing q /L by the deep percolation rate,.i_p, we get

$$L^2 = \left(\frac{4K}{i_p}\right)\left(b^2 - a^2\right)$$

The term ($b^2 - a^2$) may be expressed as follows

$$(b^2 - a^2) = (b + a)(b - a) = (a + hm + a)\ (a + hm - a) = (2a + hm)(hm)$$

Then Equation of L^2 can be rewritten as follows

$$L^2 = \left(\frac{4K}{i_p}\right) = (2a + hm)\,hm$$

$$L^2 = \frac{8K(a\,hm)}{i_p} + \frac{4K(a\,hm^2)}{i_p}$$

Hooghoudt's Equation

Hooghoudt's drainage equation (Hooghoudt, 1940) gives a mathematical relation of the parameters involved in the subsurface drainage of flat land by a system of horizontal and parallel ditches or pipe drains without entrance resistance, placed at equal depth and subject to a steady recharge evenly distributed over the area.

Assumptions

1. The soil is homogeneous with hydraulic conductivity, K;
2. The drains are equally spaced at a distance of L;

3. The hydraulic gradient at any point in the soil profile is equal to the slope of the water table above the point, i = dy/dx;
4. Darcy's law is valid for the flow of water through the soil profile;
5. There is an impermeable layer underlying the drain at a depth, D
6. The rate of replenishment of the water table by irrigation or rainfall is R;
7. The origin of the coordinates is assumed to be on the impermeable layer below the centre of one of the drains.

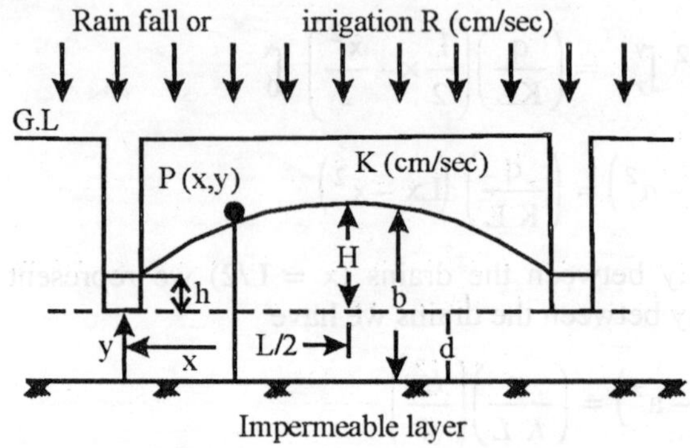

Figure 5.2. Definition sketch for Hooghoudt's equation

Considering a vertical plane through the centre line between the drains it can be seen that all the water entering the right side of plane will go the right drain and all the water entering left side will go to the left drain. Considering a unit thickness of the soil (perpendicular to the plane of the paper) the quantity of water passing through any plane at {x, y} per unit time can be given by :

$$q_x = \left(\frac{L}{2} - x\right) R$$

Now as per Darcy's law,

$$q_x = Ky\left(\frac{dy}{dx}\right)$$

Now equate above equations, we get,

$$\left(\frac{L}{2} - x\right) R = K y \left(\frac{dy}{dx}\right)$$

or

$$\frac{RL}{2}\, dx - R\, x\, dx = K\, y\, dy$$

Integrating the above equation, we get,

$$\left(\frac{RL}{2}\right) x - \frac{R\, x^2}{2} = \left(\frac{K\, y^2}{2}\right) \tag{A}$$

Now substitute the boundary conditions as

x = 0, y = h+d

x = L/2, y = H +d

We get, the equation as below:

$$L^2 = \frac{4\,K\,(H^2 - h^2 + 2\,dH - 2dh)}{R}$$

Where,

S = Spacing between two drains, m

K = Hydraulic conductivity of the soil, m/day

H = Depth of water level at mid point, m

h = Depth of water level in the drain, m

R =Rate of recharge or rainfall rate per unit area, m/day

d = Depth of impervious layer below the drain, m.

Derivation of ellipse equation

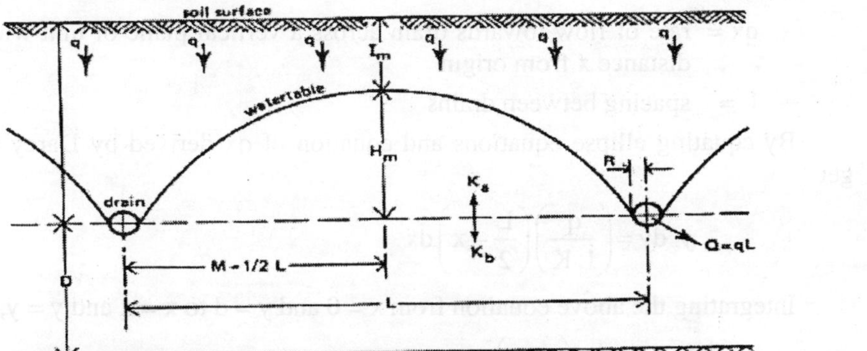

Figure 5.3. Illustration of the parameters involved in Hooghoudt's equation for Steady state condition

The above equation gives the spacing of either tile or open drains under the given set of conditions. For all practical purposes the drain is considered to be empty (h = 0). In that case the above equation reduces to

$$L^2 = \frac{4\,K\,H}{R}(2d + H)$$

The origin of the coordinate system can be shifted to the midpoint between the drains.

For this condition, the value of x substituted by $(\frac{L}{2} - x)$, we get

$$\frac{RL^2}{8} - \frac{R\,x^2}{2} = \frac{K\,y^2}{2}$$

Rearranging the terms, we can write.

$$\frac{y^2}{\left(\frac{RL^2}{4\,K}\right)} + \frac{x^2}{\left(\frac{y^2}{4}\right)} = 1$$

This equation is the equation of an ellipse.

Another method to derive this relationship in terms of total flow rate (q) into the drain per unit length and height of water table (b) from the impermeable layer, we can approach in similar fashion.

Thus, assuming that flow is removed from the drains on both side equally, we can write

$$qx = \left(\frac{\frac{L}{2} - X}{\left(\frac{L}{2}\right)}\right)\frac{q}{2}$$

where,

qx = rate of flow towards drain across a vertical plane of unit width at any distance x from origin

L = spacing between drains

By equating ellipse equations and equation of qx derived by Darcy's law, we get

$$y\,dy = \left(\frac{q}{L\,K}\right)\left(\frac{L}{2} - x\right)dx$$

Integrating the above equation from x = 0 and y = d to x = x and y = y, we get

$$\left(y^2 - d^2\right) = \left(\frac{q}{KL}\right)(Lx - x^2)$$

This equation is identical to ellipse equation and substituting x = L/2 and y = b for the mid point, we obtain,

$$L^2 = \frac{4\,K\,(b^2 - d^2)}{q}$$

But q = L i , where i= drainage rate, hence

$$L = \left[\frac{4\,K\,(b^2 - de^2)}{i}\right]^{1/2}$$

Where, *de* is the **equivalent depth** of water conducting layer below the drain.

Modifications to Hooghoudt's Equation

When deep trench drains reaching down to the impervious layer, the Ellipse equation is better suited for the estimation of drain spacing. Figure shows;

(a) a typical drainage flow pattern in the real system, and

(b) the idealized equivalent adopted for the purpose of analysis

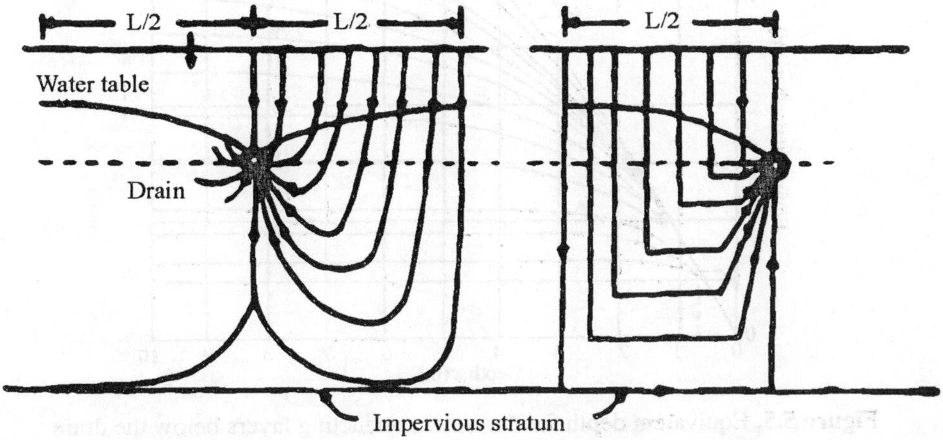

Figure 5.4. Actual and idealized patter of flow towards a subsurface drain

The recharge water first percolates vertically downwards through the unsaturated top soil to reach the water table in the idealized picture. The actual flow lines are characterized by curvatures. In the saturated zone below the water table the flow continues downward, but it is assumed to turn horizontal towards the drain. Streamlines farther from the drain first travel down vertically and then turn horizontal as indicated in the figure. Thus, near the drain the flowlines converge to feed the drain as shown. In all these three reaches, namely vertical travel, horizontal travel and inclined or radial travel towards the drain, there are energy losses occurring.

In the derivation of the Ellipse equation, we neglect the loss due to the convergence of the stream lines towards the drain, which acts more like a point - sink. If the drainage occurs into a deep open ditch which extends down to the impervious layer, the Ellipse equation would be a good approximation. This is nearly the case when the spacing between the drains, above the impervious layer, is considerably large. Over a considerable range of interest in practical cases, we find the impervious layer hydraulically too close to the drains. In such cases, we should find it necessary to use an equivalent depth 'de' in place of 'a in the Hooghoudt equation. In that case the depth tb' to be made use of in that equation would also be based on the value of de, such that h = de + hm where in hm, is the maximum permissible height of the water table above the center of the drain. Van Schilfgaarde prepared, in 1963, a chart relating the equivalent depth de and the actual depth of the impervious stratum below the ground level d, with drain spacing as a parameter, for a specified drain diameter of 127 mm. It may be noted that this chart covers a range of depth from 0 to 10 m, spacing from 6 to 91 m with the equivalent depth lying in the range of 0 to 3 m.

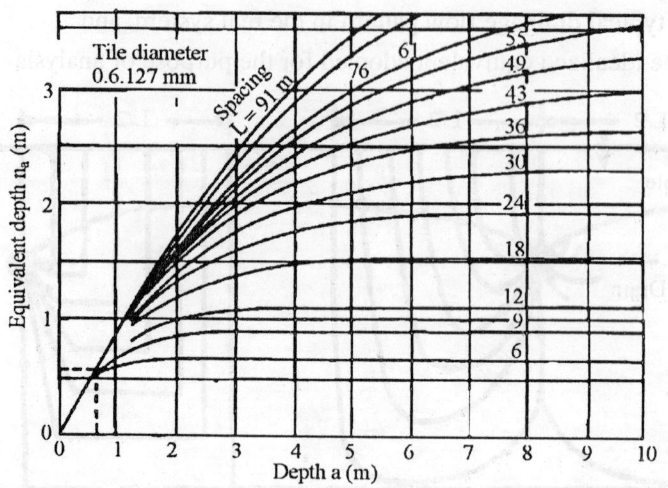

Figure 5.5. Equivalent depth for the water conducting layers below the drain

Table 5.2. Equivalent Depth 'd_e' for Commonly Used Pipe Drains (For drains with diameter = 80 to 200 mm)

Actual Depth, (a)	Spacing, L (m)										
	5	10	20	30	40	50	80	100	150	200	250
0.5	0.5	0.5	0.5	0.5							
1.0	0.7	0.8	0.9	1.0	1.0	1.0	1.0	1.0	1.0	1.0	1.0
1.5		1.0	1.2	1.3	1.4	1.4					
2.0		1.1	1.4	1.6	1.7	1.7	1.8	1.9	1.9	1.9	2.0
3.0			1.7	2.0	2.2	2.3	2.5	2.6	2.7	2.8	2.9
4.0			1.8	2.2	2.5	2.7	3.1	3.3	3.5	3,6	3.7
5.0			1.9	2.4	2.8	3.0	3.6	3.8	4.1	4.3	4.5
6.0				2.5	3.1	3.3	4.0	4.3	4.7	5.0	5.2
7.0				2.6	3.2	3.5	4.3	4.6	5.2	5.6	5.8
8.0				2.6	3.2	3.6	4.5	5.0	5.7	6.2	6.5
9.0					3.3	3.7	4.7	5.3	6.1	6.7	7.0
10.0						3.8	4.9	5.5	6.5	7.1	7.6
15.0							5.4	6.3	7.8	8.9	10.7
20.0							5.6	6.6	8.6	10.0	11.1
25.0							5.8	6.8	10.0	10.7	12.1
30.0									10.3	11.3	12.9
40.0										11.8	13.8
50.0										12.1	14.3
60.0											14.6
∞	0.7	1.2	1.9	2.6	3.3	3.9	5.8	6.8	10.6	12.2	14.7

The most widely known form of Hooghoudt's equation was presented by Wesseling (1972). In a slightly modified form, it reads:

$$qL = (8\ Hm\ /L\ )(Kb \times De + Ka \times Ha)$$

where, q is the steady recharge of water percolating to the water table equal to the drain discharge (m/day or m/hr), L is the drain spacing (m), Hm is the height of the water table mid- way between drains, taken with respect to the centre of the drain (m), Kb is the hydraulic conductivity of the soil below drain level (m/day or m/hr), Ka is the hydraulic conductivity of the soil above drain level (m/day or m/hr), De is Hooghoudt's equivalent depth to the impermeable layer below drain level, and Ha=Hm/2 is the average height of the water table above drain level.

The equivalent depth De depends on the depth D of the impermeable layer below the drains as follows:

If $D<R$: $De = D$

If $R<D<L/4$: $De = D.L/\{(L-D2)+8D.L.\ln(D/R)\}$

If $D>L/4$: $De = L/8\ln(L/R)$

where R is the drain radius (m). The two expressions yield only a very small difference in the values of De. If the drains are open ditches instead of buried pipes, the above equations are applicable with an equivalent radius calculated as $R=W/\pi$, where W is the wetted perimeter of the ditch.

Drainage of Sloping Land with Entrance Resistance

Figure illustrates the parameters involved in the drainage of sloping land with entrance resistance. The symbols used in the figure are defined as follows:

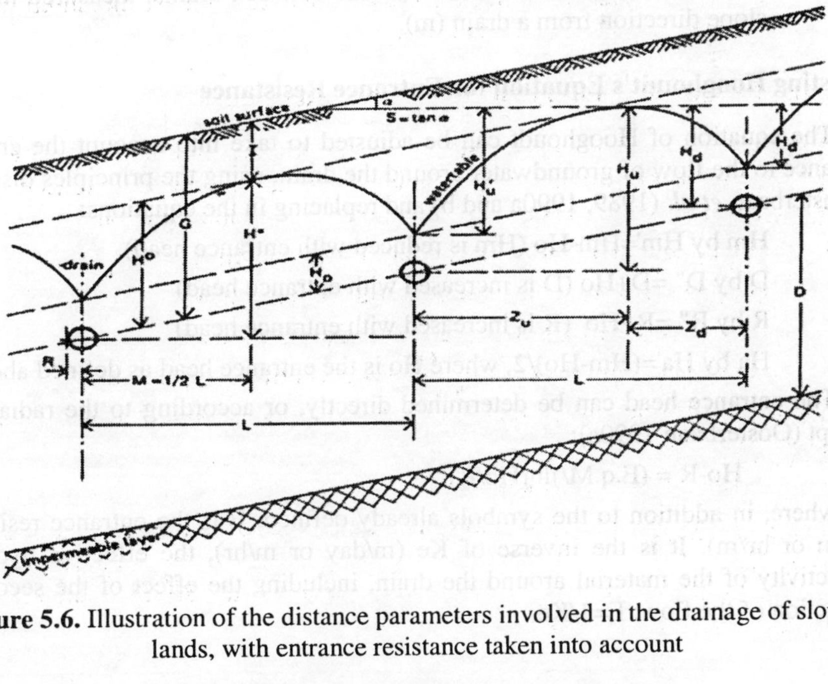

Figure 5.6. Illustration of the distance parameters involved in the drainage of sloping lands, with entrance resistance taken into account

Where,

- D is the depth of an impermeable layer below the drain centre (m);
- G is the depth of the drain centre below the soil surface (m);
- Ho is the entrance head, equal to the height of the water table just above the drain, measured from the centre of the drain (m);
- Hu is the height of the water table at the water divide between drains, measured in upslope direction from the centre of a drain (m);
- Hu'=Hu-Ho (Hu reduced with entrance head);
- Hd is the height of the water table at the water divide between drains, measured in down slope direction from the centre of the drain (m);
- Hd'=Hd-Ho (Hd reduced with entrance head);
- H* is the height of the water table midway between drains, measured from the centre of the nearest down slope drain (m);
- Hgr is the height of the water table midway between drains, measured from the sloping line through the centres of the drains (m);
- I is the depth of the water table below the soil surface midway between drains (m);
- L is the drain spacing (m);
- M=L/2 (half the drain spacing);
- S is the slope of the land (m/m);
- Zu is the distance to the water divide between drains, measured in upslope direction from a drain (m);
- Zd is the distance to the water divide between drains, measured in down slope direction from a drain (m).

Adjusting Hooghoudt's Equation for Entrance Resistance

The equation of Hooghoudt can be adjusted to take into account the entrance resistance to the flow of groundwater around the drain, using the principles discussed by Oosterbaan *et al.* (1989, 1990a and b) and replacing in the equations:

- Hm by Hm'=Hm-Ho (Hm is reduced with entrance head)
- D by D" =D+Ho (D is increased with entrance head)
- R by R" =R+Ho (R is increased with entrance head)
- Ha by Ha'=(Hm-Ho)/2, where Ho is the entrance head as defined above.

The entrance head can be determined directly, or according to the radial flow concept (Oosterbaan 1990a):

$$Ho\text{-}R = (E.q.M/)\ln(Ho/R)$$

where, in addition to the symbols already defined, E is the entrance resistance (day/m or hr/m). It is the inverse of Ke (m/day or m/hr), the effective hydraulic conductivity of the material around the drain, including the effect of the secondary contraction of the flow: E=1/Ke.

To take E or Ke into account in the finite element method, Fipps and Skaggs, 1989 used an equivalent radius r=0.01m of the drain instead of the real radius R=0.05m. In this case, the Equation 3 for entrance head reads

$$Ho - r = (q.M/K)\ln(Ho/r)$$

In the example provided by Fipps and Skaggs, we have the following data: q=0.0022m/hr, K=0.158m/hr, M=L/2=15m, and Ho=0.2m. The last value was measured from a figure in their article that is reproduced in Figure. Thus we find from Equation that Ho=0.209m, which is close to the given value Ho=0.2m.

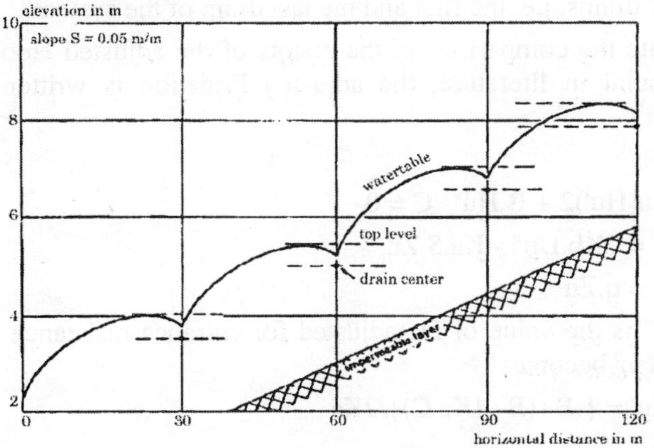

Figure 5.7. The shape of the water table obtained with the finite element method as published by Fipps and Skaggs (1989)

Adjusting Hooghoudt's Equation for Sloping Land

In Hooghoudt's drainage equation, the water divide between the drains, and therefore the zone of influence of a drain, is found at a distance M=L/2 from the drain, i.e. midway between the drains. In sloping land, with the drains placed along the contours, the zone of influence in upslope direction (Zu) is larger than in down slope direction (Zd). This implies that the water divide is not midway between the drains. However, Hooghoudt's equation can be adjusted to sloping land by using the principles introduced by Oosterbaan (1975) by which L is either replaced by 2Zu or by 2Zd, so that Equation is changed into a set of two equations.

The height (Hm) of the water table midway between the drains above drain level is then replaced by the height Hu of the water table at Zu (Hu>Hm) and by the height Hd at Zd (Hd<Hm) respectively.

Further, the average height Ha of the water table between the drains above drain level is replaced by half the height of the water table at the water divide (i.e. at Zu and Zd respectively) above the sloping line through the centres of the drains, giving respectively Hau=(Hu-S.Zu)/2 and Had=(Hd+S.Zd)/2, where S represents the slope (m/m).

Finally, the drain radius R is replaced by Ru=R.Zu/M and Rd=R.Zd/M respectively, because the water from the upstream side enters the drain over a proportionally larger part of its circumference. Now, the following relations exist

Zu+Zd = 2M

Hu-Hd = S(Zu+Zd)

Q = q(Zu+Zd)

where, Q is the steady discharge from the drain in m/day per m length of drain (m/day).

The general condition for the application of the adjustments is Zu<L and Zd>0. With extremely steep slopes (S >10%), these conditions are perhaps not met, but then drainage is not a realistic proposition anyway. Further, the adjusted equations are not valid for border drains, i.e. the first and the last drain of the system.

To facilitate the comparison of the results of the adjusted Hooghoudt equation with results found in literature, the adjusted Equation is written as a quadratic equation in Hu':

with

Ka(Hu')2 + B.Hu' - C = 0

B = 2Kb.Du" - Ka.S.Zu

C = q.Zu

where Du" is the value of De adjusted for entrance resistance and slope. Thus the solution of Hu' becomes

Hu' = {-B+(B+4Ka.C)}/2Ka

Spacing of the drains in a layered media

The original formula developed by Hooghoudt, considered the possibility of soil above the impermeable base being stratified in two layers: the one above the drain center line being of permeability Ka, and that below the drain base being of permeability Kb. The figure demonstrates such a condition.

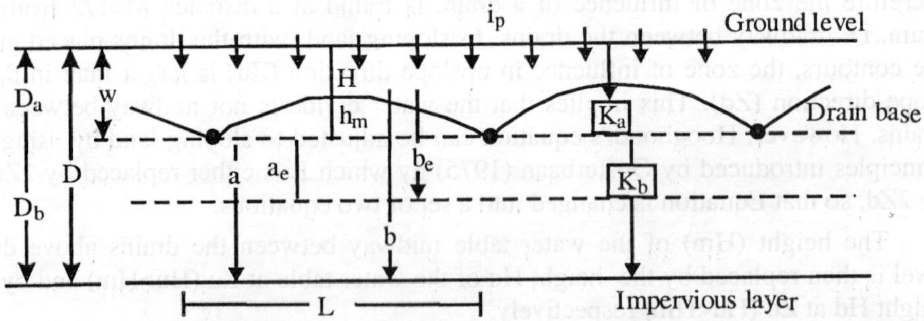

Figure 5.8. Steady flow into Drains in Layered Media

Then the following equation can be shown to be applicable

$$L^2 = \frac{4\,K_a\,hm^2}{i_p} + \frac{8\,K_b\,hm\,d_e}{i_p}$$

Where

Ka = hydraulic conductivity above the drain (m/day),

Kb = hydraulic conductivity below the drain (m/day),

hm = height of the water table above the drain level midway between the drains (m),

i_p = seepage rate or deep percolation rate which equals the drain discharge rate (m/day), and

d_e = Effective depth of layer below the drain.

It is to be noted that *de* depends on the depth of the impervious layer below the drain base *a*, and the spacing between drains, L and also the diameter of the pipe as well its wetted perimeter u. Thus the procedure for determination of L turns out to be an iterative process. The nomograph prepared by Van Beer (Figure) helps in avoiding this trial-and-error process

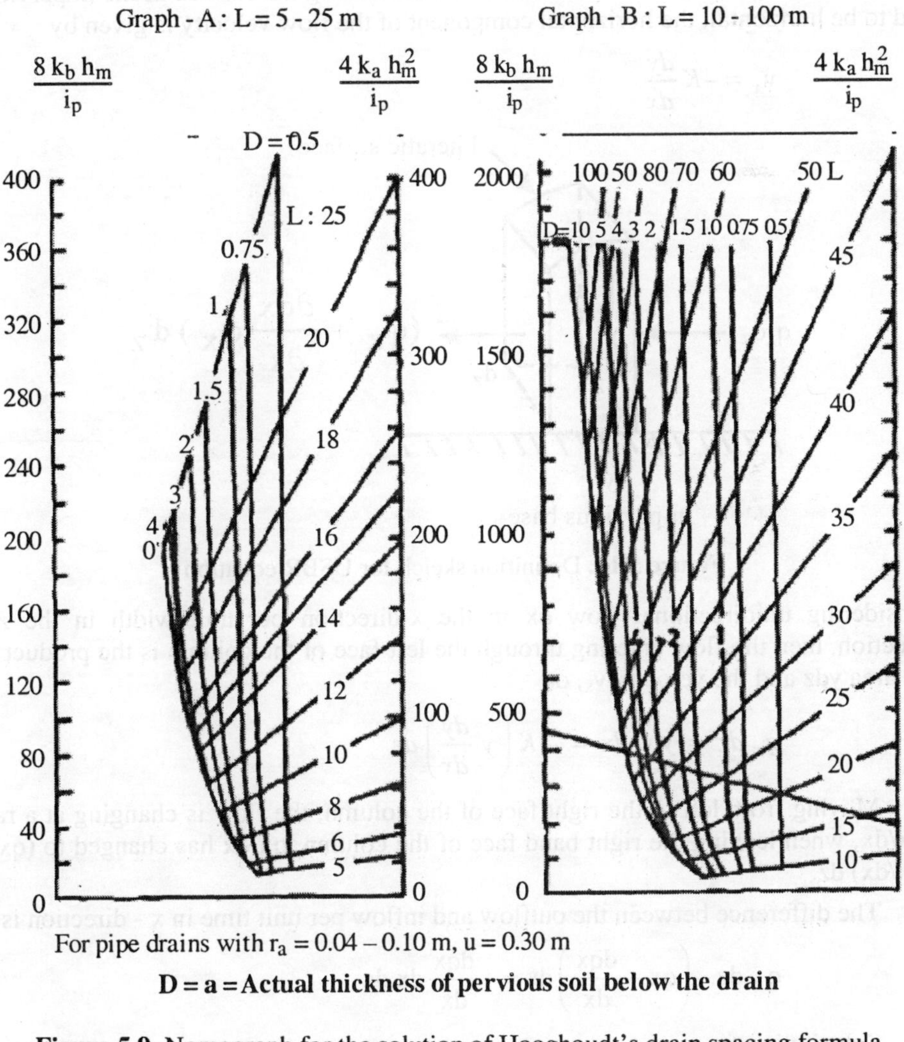

Figure 5.9. Nomograph for the solution of Hooghoudt's drain spacing formula

5.7 UNSTEADY STATE CONDITION

To find out the spacing between two subsurface drains under unsteady state condition, different equations are proposed for different situation like van Schilfgaarde (1963). As stated earlier Hooghoudt's formula considers the water table to be in equilibrium with rainfall or irrigation water. It is, therefore, referred to as a steady state equation. When the position of water table with respect to time is considered the resulting condition is known as the transient or unsteady state. The USBR equation is for the transient state of the water table. In deriving this equation all the assumptions as in case of deriving the Hooghoudt's equation are made.

U.S. Bureau of Reclamation (USBR) Formula

Consider an elemental volume of dx dy dz. Taking the surface of the impervious bed to be horizontal, the horizontal component of the flow velocity is given by

$$v_x = -K\frac{dy}{dx}$$

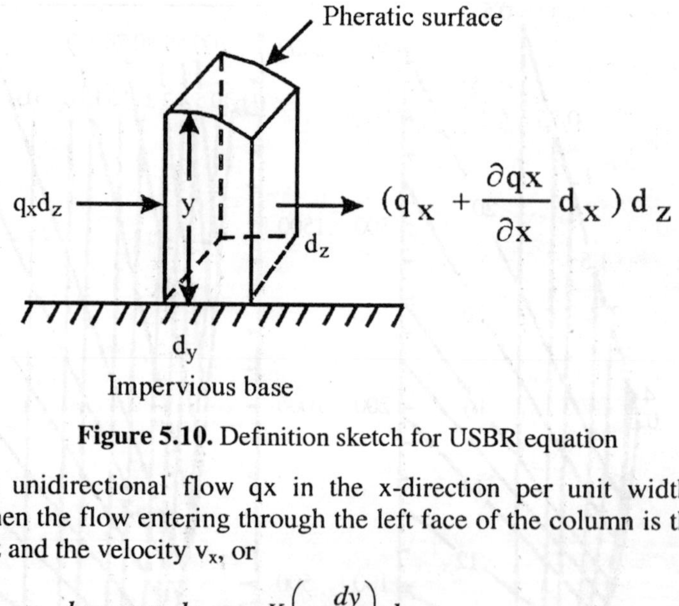

Figure 5.10. Definition sketch for USBR equation

considering unidirectional flow qx in the x-direction per unit width in the z - direction, then the flow entering through the left face of the column is the product of the area ydz and the velocity v_x, or

$$q_x\, dz = y\, v_x dx = -K\left(y\,\frac{dy}{dx}\right) dz$$

Moving from left to the right face of the column, the flow is changing at a rate dqx/dx, when leaving the right band face of the column, qx dx has changed to (qx + dqx/dx) dz.

The difference between the outflow and inflow per unit time in x - direction is

$$q_x\, dz - \left(qx + \frac{dqx}{dx}\right) dx = -\frac{dqx}{dx}\, dx\, dz$$

Substituting for

$$q_x = K\frac{d}{dx}\left(y\frac{dy}{dx}\right)dx\,dz$$

If in time Δt, there is a change in height, Δy and if f be the drainable porosity, we have

$$f\,\Delta y\;dx\,dz = K\frac{d}{dx}\left(y\frac{dy}{dx}\right)\Delta t$$

or

$$f\frac{\Delta y}{\Delta}\;\Delta t = K\frac{d}{dx}\left(y\frac{dy}{dx}\right)$$

Putting the limits i.e. when Δy =0 and Δt = 0 and considering x direction only the above is written as

$$f\frac{\Delta y}{\Delta t} = K\frac{d}{dx}\left(y\frac{dy}{dx}\right) = K\frac{d}{dx}\left[y\frac{d^2y}{dx^2} + \left(\frac{dy}{dx}\right)^2\right]$$

If y is large as compared with changes in y, we may consider y a constant with the mean value D and neglecting second order terms, we have

$$\frac{dy}{dt} = \left(\frac{KD\,d^2y}{f\,dx^2}\right)$$

where,

y = height of water table,

K = hydraulic conductivity,

D = average thickness of the aquifer from the impermeable layer to the water table, and

f = drainable porosity in per cent by volume (equivalent to specific yield).

Now consider the following figure, wherein $D = d + m_o/2$. Thus above equation with partial derivative terms can be written as equation below:

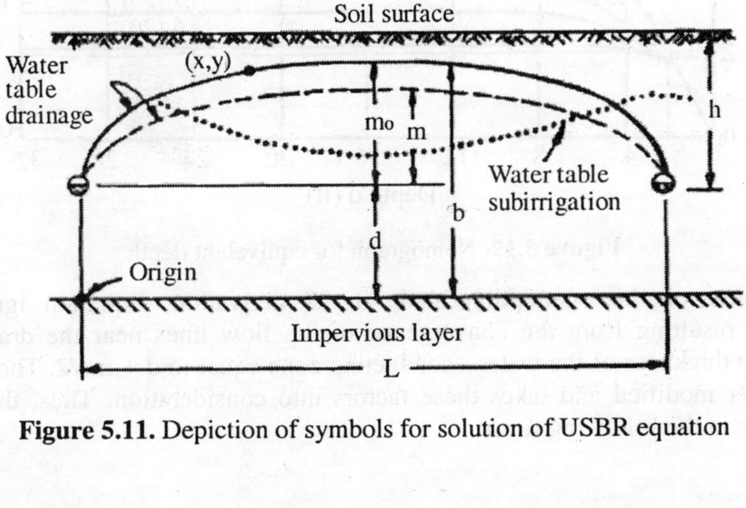

Figure 5.11. Depiction of symbols for solution of USBR equation

For study of non-steady state conditions' the Dupuit-Forchheimer theory is applicable. Here it is assumed that flow is horizontal and the velocity is proportional to the slope of the free water surface. The equation may be written as

$$\frac{d^2 y}{dx^2} = \frac{f}{\Delta K\left(d + \frac{mo}{2}\right)} \frac{dy}{dt}$$

The equation is identical to heat flow equation. By assigning appropriate boundary conditions equation can be solved in terms of a sine series retaining only the first term. The spacing equation as derived by Glover using the symbols as described in Figure is given by

$$L = \pi \left[\frac{K t \left(d + \frac{mo}{2}\right)}{f \ln \left[\left(\frac{4}{\pi}\right)\left(\frac{mo}{m}\right)\right]} \right]^{\frac{1}{2}}$$

This equation is also known as Glover's Equation for tile spacing.

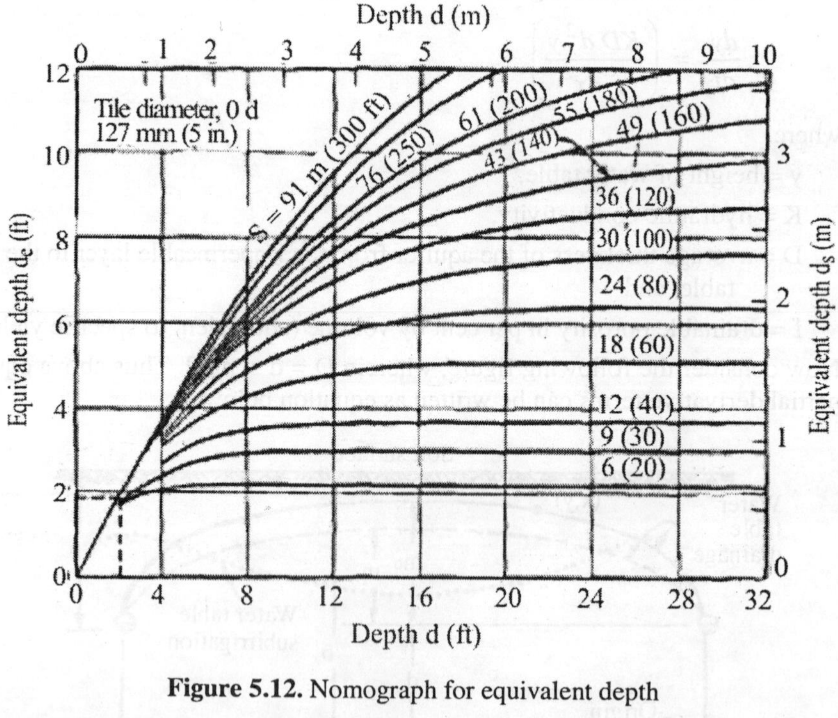

Figure 5.12. Nomograph for equivalent depth

Because of the assumption made in its derivation, Equation ignores the resistance resulting from the convergence of the flow lines near the drain and is based on a thickness of the water - conducting zone equal to d + mo/2. The equation was further modified and takes these factors into consideration. Thus, the present equation is in the following form:

$$L = \frac{9K\,t\,de}{f\left[ln\,[mo\,(2de + m)\,] - ln\,[m\,(2\,de + mo)\,]\right]}$$

where,

de = Equivalent depth from Figure,

m =Height of water table above the centre of the drain at mid plane after time t,

mo = Initial height of water table,

f = Drainable Porosity in fraction, and

t = Time for water table to drop from mo to m.

The remaining symbols carry their usual meaning.

The Equation is recommended for design purposes only as it does not yield a practical solution where de = 0

Example

Calculate the drain spacing for a falling water table assuming the value of K as 35 cm/day, f = 0.02, drain depth to centre line = 0.9 m, mo = 0.9 m, m = 0.6m, t=l day for the water to drop from the soil surface to 30 cm below, and depth of impervious layer below drains d = 0.6 m. Also calculate drain depth for a 100 mm diameter tile if initial water table is at soil surface.

Ernst Equation

The Ernst Equation is applicable to any type of two-layered soil profile. It has the advantage over the Hooghoudt equation that the interface between the two layers can be either above or below drain level. It is especially useful when the top layer has a considerably lower hydraulic conductivity than the bottom layer. To obtain a generally applicable solution for soil profiles consisting of layers with different hydraulic conductivities, Ernst divided the flow to the drains into a vertical, a horizontal, and a radial component. Consequently, the total available head (h) can be divided into a head loss caused by the vertical flow, the horizontal flow and the radial flow.

$$h = R\frac{D_V}{K_V} + R\frac{L^2}{8\,KD} + R\frac{L}{\pi K_2}\,ln\frac{a\,D_2}{u}$$

Where,

R = Rate of recharge or rainfall, m/day

D_V = Thickness of layer in which vertical flow considered, m

KD = The sum of the product of permeability (K) and thickness (D) for horizontal flow components.

K_V = Hydraulic conductivity for vertical four m/day

K_2 = Hydraulic conductivity

D_2 = Thickness of layer in which radial flow considered, m

a = Geomatric factor for radial flow

m = Wetted section of the drain, m.

5.8 DRAINAGE COEFFICIENT

The *rate of drainage* is a key factor in establishing the needed capacity of a drainage system. This rate, expressed as the depth in centimetres of water drained off from a given area in 24 hours, is called the *drainage coefficient* or *drainage design rate.* It is the design value at which water is to be removed from a drainage area. It may also be expressed in terms of flow rate per unit of area, as cubic metres per square kilometre per 24 hours, or in terms of the flow rate per unit of area which varies with the size of the area. Where the annual rainfall varies from 750 to 1500 mm, the usually adopted value of drainage coefficient varies from 10 and 13 mm tin 24 hours) under average soil conditions For organic soils or for high value crops, this value is usually increased by 30 to 50%. Where surface runoff is to be carried by drain tiles (through surface inlets), the rate is usually doubled, i.e. 20 to 25 mm.

Factors affecting drainage coefficient

The coefficient also varies with the geographic location, land use, and size of area to be drained. While considering the drainage requirement of humid areas, the rainfall amount, frequency, intensity, and duration; other climatic factors; soil permeability; crops to be grown; and the size of the area to be drained are considered. In arid and semi-arid irrigated areas, the irrigation system and field application losses, rather than the rainfall, are the principal factors on which the drainage coefficients are based. Drainage coefficients are selected with respect to the degree of protection to be provided for various crops. For average small drainage projects the drainage coefficient would range from 6 to 25 mm. The value of the drainage coefficient for a particular-area is usually determined on the basis of past experience with similar soil, crop and climatic conditions. Rainfall is the most important factor influencing the value of the drainage coefficient in places where drainage is needed to remove rainfall-runoff.

Determination of Drainage Coefficient

The common procedure of determination of drainage coefficient is based on calculating an appropriate rate of removing the runoff from agricultural lands that has been estimated any other method. A design drainage coefficient is obtained by considering the runoff from the design rainfall. A design rainfall may be considered as the rainfall of a given duration with a given recurrence interval. For example, one may determine a drainage coefficient for a 5-year 1 -day rainfall. There are two approaches to determine the drainage coefficient. In one of the approaches, the estimated runoff from the design rainfall is distributed over time by following a certain calculation procedure. The result is a hydrograph of flow with a peak. Under this approach, design drainage rate can also be estimated by using some empirical relations. The drainage channel capacity is determined to carry the peak flow as obtained from the above analysis. Such a procedure will result in too large dimensions of drainage channel cross-section which is not required and will be expensive. In the other approach, for designing the capacity of the drainage channel, a certain time is assumed as the time within which the excess runoff has to be removed.

A safe excess water removal period may be considered as 2 days. This is based on the assumption that if the excess water is removed from the field within 2 days, there may not be any irreversible adverse effect on the plant and hence, on the crop yield. It would appear appropriate to consider a 2-day rainfall of a desired recurrence interval as the design rainfall, determine the corresponding direct runoff and divide this by the stipulated period of excess water removal to get the design drainage coefficient. Both the above approaches are the ways of indirect estimation of drainage coefficient. An economic drainage system is one which may lead to the desired benefit from drainage (e.g., production at a pre-determined level) at a minimum cost. It is often difficult to quantify the benefits from drainage other than the increase in crop production due to good drainage. Such benefit components are better trafficability, more field working days, lesser insect-pest attack and the like. These act as co-variables in increasing the crop yield.

USSCS (United States Soil Conservation Service) (1973) has recommended the following relationship for estimating drainage coefficient from irrigated areas.

Drainage coefficient, DC = (P + c) i / F

Where,

DC = Drainage coefficient, mm/day

P = Deep percolation from irrigation, including leaching requirement as decimal of the per cent of the flow.

c = canal or water seepage losses, as a decimal of per cent of the flow

i = irrigation application, mm

F = Irrigation interval, days.

Determination of Drainage coefficient using rainfall data

Collect the maximum 1-day rainfall of year records for at least last 35 years. Arrange the data in descending order and assigned order m =1 to the highest value and m = n to the lowest value. Find return period using T = (n+1)/m. Draw the frequency curve for 1-day maximum rainfall by taking log value of T on x-axis and rainfall value on y-axis. Find the maximum 1-day rainfall for return period of 15 years. Find excess rainfall using standard method

$$Pe = P_{total} - (I \times 24)$$

Drainage coefficient = Pe / (1+t/24)

where, t = submergence resistance period, day.

The design capacity of tile drains is based essentially on the drainage coefficient, the drainage coefficient is the depth of water to be removed from the drained area in one day (24 hours). In humid areas the drainage coefficient depends largely on rainfall. Since the rainfall is not constant and varies greatly in amount and intensity during the rainy season, it is difficult to estimate a reliable value of drainage coefficient under humid conditions. However, several workers have suggested guidelines for estimating the drainage coefficient Where the annual rainfall varies from 750 to 1500 mm, the usually adopted value of drainage coefficient varies from 10 and 13 mm tin 24 hours) under average soil conditions For organic soils or for high value crops, this value is usually increased by 30 to 50%. Where surface runoff is to be carried by drain tiles (through surface inlets), the rate is usually doubled, i.e. 20 to 25 mm.

In irrigated areas, the discharges from drains may be expected to vary from 10 to 50% of the water applied. Since not all of the area is irrigated at the same time, the design flow is estimated from the area actually being irrigated. Seepage from outside the area and low flow from drains in land previously irrigated are to be considered in selecting the value of the drainage coefficient. The drainage coefficient depends on the depth of irrigation, the method of application, the leaching requirement and soil characteristics.

5.9 HYDRAULIC DESIGN OF SUBSURFACE DRAINAGE SYSTEMS/ TILE DRAINS

There are three ways to estimate the amount of water to be removed by a subsurface drainage system.

a. To remove the excess water applied by irrigation.

b. To remove the excess rainfall (in excess of the moisture holding capacity of the soil)

c. To meet the leaching requirement in situations of salt affected soils or while irrigating with saline water.

The capacity and length of the drain pipes for various diameters and slopes are considered and the maximum area drained by field drains and collectors are determined. Then this is used in combination with the drain spacing calculations, to determine the layout of the pipe drainage system. The amount of water to be conveyed by a pipe drain follows from the design drainage coefficient and the area served by the pipe.

Capacity of tile drains

After the spacing of the tile drains is determined, the discharge through the tile drains can be calculated either by using the Hooghondt's approach for steady state conditions or the USBR approach for the unsteady state condition. In Hooghondt's approach, the maximum rate of flow per unit length of the drain is given by

$$Q = S\,R / 2$$

In USBR approach, the capacity of the tile drains can be found out by the equation below:

$$Q = \frac{2\pi h k d}{S}$$

Where

Q = discharge of drain, cum/sec

h = height of the water table, m

k = coefficient of permeability, m/day

d = diameter of drain, cm

S = drain spacing, m

The design flow of tile drains is estimated under humid and arid condition using different approaches. Under humid conditions, the main purpose of the tile drains is to remove the excess rainfall appearing as sub-surface flow. The concept of drainage coefficient is used for determining the capacity of the field drains. The following table gives the values of the drainage coefficient for field crops. These values vary depending upon the local soil, crop and surface drainage conditions. Some experience is necessary for deciding the value of the drainage coefficient.

Table 5.3. Drainage Coefficient for Tile Drains in Humid Regions

Surface Drainage Condition	**Drainage Coefficient (cm/day)**	
	Mineral Soils	**Organic Soils**
(a) Normal	1.0 - 1.3	1.3 - 1.9
(b) With blind inlets	1.3 -1.9	1.3 - 2.5
(c) With surface inlets	1.9 - 2.5	1.9 - 3.7

In determining the capacity of the tile line, the area actually to be drained by that particular line is to be considered. Additional allowance is to be made where there are surface inlets into the tile line or side hill seepage. The Manning's formula is used to find out the hydraulic capacity of drains. The slope of the drain pipe and the roughness coefficient, n are to be known in addition to the discharge coefficient. Minimum longitudinal gradients, where siltation is not a hazard is 0.1% for 10 cm diameter tile and 0.05% for a 15 cm diameter tile.

Size of tile drain

The size of the tile drains is determined using the maximum expected flow and the grade. Manning's formula is used for the design.

$$Q = A\,v$$

$$Q = \frac{\pi}{4} d^2 \frac{1}{n} \left[\frac{\frac{\pi}{4} d^2}{\pi d}\right]^{\frac{2}{3}} S^{\frac{1}{2}}$$

$$d = 51.7\left(D_c \times A \times n\right)^{0.375} \times S^{-0.1875}$$

Where,

d = inside diameter of drain, mm.

D_c = derange coefficient, mm/day

A = drainage area, ha

n = Manning's roughness coefficient

S = slop of drain in m/m.

Diameter of the drain pipe

A minimum diameter of 10 cm is generally used for clay and concrete pipes. Collector drain pipes may be of much larger diameter, up to 75 cm. Only when the laterals are more than 150 m long it is necessary to compute the size of the tile needed, for the standard sizes of 10 cm and 12.5 cm have more than adequate capacity for laterals of shorter length.

Grade of tile drain

Tile grades should be such that no sedimentation occurs in the tiles. A minimum velocity of 45 cm/second is generally recommended. A desirable working grade is 0.2 per cent while higher grades can be adopted. The desirable grade for lateral is 0.2% and for main drain line the grade very up to 2-3%.

Table 5.4. Minimum Grades for Pipe Drains

Nominal Diameter (mm)	Fine Sand or Silt Not Present		Fine Sand or Silt May Enter the Drain	
	Tile	Tubing	Tile (Grade%)	Tubing
75	0.08	0.10	0.60	0.81
100	0.05	0.07	0.41	0.55
125	0.04	0.05	0.30	0.41
150	0.03	0.04	0.24	0.32

(a) Minimum cleaning velocity of 0.15 m/s

(b) Maximum cleaning velocity of 0.42 m/s

Roughness coefficient

The roughness coefficient, n in Manning's equation will vary with the type of drain pipe, their smoothness, alignment at tile joints and corrugations in tubing. The design values of n, usually adopted, are 0.013 for clay and concrete pipes and 0.015 to 0.02 for corrugated plastic tubing.

Table 5.5. Values for Manning's Coefficient 'n' for Pipe Material

Pipe material	Manning's Coefficient 'n'		
	Min	Normal	Max.
Pipes Flowing Partly Full			
corrugated metal, subdrain	0.017	0.019	0.021
corrugated metal, storm drain	0.021	0.024	0.030
Non-metal			
concrete, straight and free of debris	0.010	0.011	0.013
concrete, bends, connections, some debris	0.011	0.013	0.014
concrete, finished	0.011	0.012	0.014
concrete, steel form	0.012	0.013	0.014
concrete, smooth wood form	0.012	0.014	0.016
concrete, rough wood form	0.015	0.017	0.020
clay common drainage tile	0.011	0.013	0.017

Load on the tile drain

The Ditch Conduit Formula is used for calculating the load on pipes in narrow trenches and is given by:

$$W_c = C_d \; W \; B_d^2$$

Where,

W_c = load on the pipe

C_d = load coefficient

W = unit weight of fill material

B_d = width of trench at the top of pipeline

Tile joint spacing

The maximum permissible crack width usually recommended for silt loam soil and silt clay loam soil vary from 1 mm to 3 mm. In sandy and sandy loam soils their ends are placed as close as possible.

Perforations in drain pipe

As the number of holes is increased the relative increase in flow decreases, if there are more than 65 holes per meter. Doubling the diameter of the perforations from 6 to 12 mm at 1.2 meter depth increased the flow 68 and 46% for 13 and 33 holes. The use of envelope materials around the perforated pipe would reduce the number of perforations in order to get the desired flow. As the number of holes is increased the relative increase in flow decreases, particularly if there are more than 65 holes per meter.

5.10 ECONOMICS AND COST ANALYSIS OF DRAINAGE SYSTEM

Interest

It is a reward for having made capital available to someone who needs it.

Simple Interest

Rate of interest is imposed in certain period of time regularly on present worth value.

Compound Interest

It is the rate of interest imposed at fixed period of time regularly on present worth value and interest amount.

A. Single Payment factor

1. Single payment Compound Amount or Future Worth Factor

$$F/P = (1 + i)^n$$

2. Single Payment Present Worth Factor (Present sum that is equivalent to amount invested after n years)

$$P/F = 1 / (1 + i)^n$$

B. Uniform Annual Series Factors (A/F)

1. U.A.S. Sinking fund factor (A/F) (What has to be paid every year to amount after n years)

$$A/F = i / [(1 + i)^n - 1]$$

2. U.A.S. Capital Recovery factor (A/P) (Which has to be paid every year to be equivalent to amount invested now).

$$A/P = [i (1 + i)^n] / [(1 + i)^n - 1]$$

3. U.A.S. Future worth factor (F/A) (The sum to which amount per year will accumulate after n years)

$$F/A = [(1 + i)^n - 1] / i$$

4. U.A.S. Present worth factor (P/A) (The present sum that is equivalent to amount per year during n years)

$$P/A = [(1 + i)^n - 1] / [i (1 + i)^n]$$

C. Gradient Series

1. Present Worth Factor

$$P/G = [(1 + i)^{n+1} - (n i + i + 1)] / [i^2 (1 + i)^n]$$

2. Future Worth Factor

$$F/G = [(1 + i)^{n+1} - (n i + i + 1)] / i^2$$

D. Uniform Per centage Gradient Series

1. Present worth factor

$$Pj = [(1 + i)^n - (1 + j)^n] / [(i - j) (1 + i)^n]$$

2. Future worth factor

$$Fj = [(1 + i)^n - (1 + j)^n] / [(i - j)]$$

For the comparison among various agricultural drainage projects, Following criteria are considered:

1. Profit = Benefit - Cost

Input and output amount are not compare, only difference is considered. For limited fund.

2. B/C ratio = Benefit/Cost

The ratio of output to input and do not considered the amount of profit. For unlimited fund.

3. Rate of Return

Different studies carried out to know the benefits of the drainage systems in different crop system slike Schwab and *et al.* (1966) studied the crop response from tile and surface drainage. Rao *et al.* (1992) studied the effect of subirrigation by groundwater management with controlled subsurface drainage in semi-arid areas. Oosterbaan, *et al.* (1990) evaluates the crop production with the saline soil. They evaluated field data from India by segmented linear regression. Rudd and Chardon (1977) studied the effects of drainage on cane yields as measured by water table height in the Machnade Mill area.

Example

Suggest the design of the subsurface system for 100 ha waterlogged area, The length of the field is 1000 m. The slope towards the centre of the field is 0.2% and the slope along the length is 0.15%. The natural outlet is located at the end of the field. Water level in the outlet is 2.5 m below the ground surface. The EC of water is 2 dS/m. The crop tolerance is 4 dS/m. The net cutoff per irrigation is 8 cm. Hydraulic conductivity of the soil is 0.7m/day and the soil is considered homogeneous. The drainable porosity is 8%. Drain off period is 15 days. Depth of impervious layer from the ground surface is 5.5 m. Work out the quantities of different items for cost estimation. 1. Earth work 2. Sand volume of 6 cm thickness. 3. Gravel envelope material is 7.5 cm. Total length of PVC pipe of lateral and main line. 4. Laying and joining.

Solution

Steps

- Find out the Leaching Requirement using equation:

  LR = ECi/ECd

- Find out the Depth of irrigation water using equation:

  Di =ECd/(ECd-ECi) x Dc

- Find out the Depth of drain water using equation:

  Dd = Di X LR

- Using the drainable porosity find out the rise in water Table

  = Dd / μ

- Thus, the parameters can be defined as follows
- Using modified Glover-Dumm equation (Dumm, 1954), the spacing between drains can be calculated as

$$S = \pi \left(\frac{K\, d\, t}{\mu} \right)^{1/2} \left(ln\, 1.16 \frac{h_o}{h_t} \right)^{-1/2}$$

Where,

Ht = 25 cm, ho= 150 cm, K = 0.6 m/day, m = 0.08

t = 25 days, D = d + (ho/2) = 3 + (1.5/2) = 3.75 m

- Find out the number of laterals:

  = Length of lateral /Spacing between drains (S)

- Find out the total length of all laterals
- Now estimate the area to be drained by each lateral
- Estimate the drainage coefficient,

  DC = Depth of drained water/ Irrigation interval

- Find out the drainage discharge from each lateral
- The diameter of drain pipe can be determined by using Manning's formula for drain discharge, which can be given as

$$Q = \frac{1}{n} R^{\left(\frac{2}{3}\right)} i^{\left(\frac{1}{2}\right)} A$$

- Find out the diameter of collector drain
- Split the field lengthwise into required feasible sections and find out discharge through each section. Use above equation to find out the diameter of collector of each section.

Cost Estimation

- As per SOR, find out the Cost of lateral pipe, Cost of collector drain for different sections and Total cost of collector drain pipe.

  Total cost of piping = Cost of lateral pipe + Cost of collector drain pipe
- Take trapezoidal section and use following equation to find out the cross sectional area and earthwork for lateral and collector.

  $$A = bd + zd^2$$
- Volume of earth work of drain = Area x Length of lateral
- Find out the cost of earthwork of laterals and collector drain,
- Total excavation charges = Cost of excavation of lateral + Cost of excavation of collector drain
- Find out the volume and cost of sand required for sand bedding.
- Find out the volume and cost of Gravel envelop required only for laterals.
- Find out the cost of gravel material, cost of laying and fitting, cost of sand bedding and gravel envelope and cost of backfilling of collector drain.
- Volume of backfilling = Vol. of earthwork of collector drain - Volume of pipe - Volume of sand .
- Find out the cost of backfilling of collector drain.
- Volume of backfilling = Vol. of earthwork. of lateral pipe - [Vol. of pipe + Vol. of without pipe] – Vol. of sand
- Find out the cost of backfilling of lateral drain pipe.
- Find out the cost of Gypsum requirement.
- Find out the cost of placement of joint and manholes.

Cost for 100 ha

- = Cost of excavation + Cost of collector drain + Cost of lateral pipe + Cost of sand + Cost of gravel + Cost of laying and fitting' + Cost of sand bedding and gravel + Cost of backfilling + Cost of gypsum + Cost of manholes + Cost of placement of joint

- Cost for 1 ha = **Cost for 100 ha/ 100**

Profit from different crops:

- Find out the Cost of Cultivation (Rs.)
- As per the total yield, find out the Gross Income (Rs.).
- Deduct the cost from the gross income and find out net profit for different crops.
- Find out the Benefit Cost Ratio = Benefit/ Cost

Exercises

1. Design a subsurface drain system for an irrigated area to drain excess water. The impermeable medium is encountered at a depth of 900 cm from the ground surface and the root zone depth for the crop concerned is 120 cm. The medium in between has an average hydraulic conductivity of K = 300 cm /day. It is estimated that 2.5 cm, out of the irrigated water that is supplied, seeps below the root zone as deep percolation in every irrigation session. The irrigation interval is two weeks. Fix the position and estimate the spacing of drains. Also work out the spacing of drains, if the drains are set at a depth of 1.8 m below the ground surface.
2. The irrigation recharge rate of 1.3 mm/day into a soil with a hydraulic conductivity of 0.4 m/day in a cultivated area was estimated. The pervious soil layer is 5 m thick. It is desired that the water table should be at least 1.5 m below the ground level. For an assumed depth of the drain as 1.8 m below the ground level, estimate the spacing between the drain considering the following factors : (a) the actual depth to the impervious layer, and (b) the equivalent depth. The drain diameter may be assumed as 125 mm. (Van Schlifgaarde chart may be used).
3. Compute the drain spacing, for an irrigated area, assuming the depth to the centre of drain is 1.8 m and the minimum depth to the water table is 1.5 m. Given that hydraulic conductivity K= 0.5 m/day above an impervious layer at a depth of 6.7 m. The excess irrigation rate is equivalent to drainage coefficient of 1.3 mm/day. Assume equivalent depth, de = 2.9 m.
4. A drainage canal discharges 0.4 cubic metres of water per second and drains 550 hectares. What is the drainage coefficient of this land?
5. The drainage coefficient of a land is 1 cm. Calculate the capacity required at the outlet end of the drainage ditch draining a watershed of 3 sq.km.
6. Find out the size of the tile required at the end of a 500 m long tile line, if the drainage coefficient is 1 cm, grade is 0.3% and tile spacing is 50 m.
7. Find out the drain spacing for an irrigation area, assuming the depth to the center of drain is 1.8 m and the minimum depth to the water table is 1.5 m given that hydraulic conductivity k = 0.5 m/day above the impervious layer at a

depth of 6.5 m the access irrigation rate is equal to daring coefficient of 1.3 mm/day.

8. Determine the diameter of corrugated plastic tubing and the flow rate where the slope is 0.4 per cent, drainage area is 12 ha and drainage coefficient is 15 mm/day.

9. In a sub-surface drainage system, 200 m long laterals were laid out 50 in apart. The laterals have a grade of 0.3 per cent, (a) If the drainage coefficient of the area is 2 cm per day, what size of tiles would you recommend? (b) If the drainage coefficient is increased to 3 cm per day, what will be the spacing of the laterals?

10. The size of a branch pipe at a given location is to be determined, given the following data : (a) Area to be drained by the field drains = 15 ha (b) The drainage coefficient = 10 mm/day (c) Factor of safety for design discharge = 1.5 (d) The pipe used is of the corrugated variety: to be laid at a slope of 0.25 per cent.

11. Determine the size of the tile required at the end of a 500 m long tile line, if the drainage coefficient is 1 cm, grade is 0.3 per cent and die spacing is 50 m.

12. A subsurface drainage system is to be installed to control the water table under the following conditions: Drainage coefficient is 1 mm/day. The depth of water table midway between the drains is to be kept at 1.0 m below the soil surface. The drains will be installed at a depth of 2 m. The drain pipe with a radius of 0.10 m will be used. There will be constant supply of water.

13. The depth of the root zone is 1 m of a crop which is suitable for a soil. The water table depth is 0.5 m below ground surface. The hydraulic conductivity is 1 cm/day. The depth of the tile is 1.5 m below the ground surface. The soil of the filed is fill in a sand tank and compacted with same bulk density of the field soil and found that q = (50y-30y2) k, Where, q = discharge (cum/hr), k = hydraulic conductivity (1 cm/day), y=depth of water table in tank. The drainable porosity is 10%. The crop is tolerable for 3 days to water submergence.

14. The rice is irrigated with the saline water; the average salt concentration of the irrigation water during cropping period is 1.00 dS/m. The maximum salinity that can be tolerated by the crop is 3.00 dS/m. The average evapo-transpiration during the period is 5.5 mm per day. Found out the monthly leaching requirement.

15. Design the subsurface drainage system and determine the spacing between drains using - (a) Hooghoudt's formula, and (b) Van Beer nomograph for the following conditions. The impervious layer exist at 4.5 m below the ground level. There are two different layers exists, the top layer has the thickness of 1.2 m. The depth of root zone of particular crops is 0.75 m. Permeability of the top 1.2 m thick layer was 1.8 m/day for bottom layer it was 0.85 m/day. The deep percolation rate was 6 mm/day.

16. Estimate the discharge to be carried by a lateral of a subsurface drain system, which has a catchment area of 0.70 ha assigned to it. Assume the design rainfall intensity of 9 mm/day and a drain flow factor of 0.75.

CHAPTER – 6

DRAINAGE MATERIALS

6.1 DRAIN PIPES

The clay, concrete and plastic pipes give satisfactory results. Collector pipes are made of concrete or plastic. Pipes that are manufactured from the latter type of material are not yet competitive for diameters exceeding 200 mm. However, perforated corrugated plastic collectors, wrapped with a sheet envelope, may be installed comparatively easily if the surrounding soil consists of quicksand or has other quick properties. Once installed, the collector can act as a drain, cancelling the quick condition of the soil and facilitating the connection of laterals and/or the installation of manholes.

In theory, there are valid considerations to select specific types of drainpipe. In practice, selection is mostly based on cost comparison and on local availability. In addition, the following observations may be relevant (Schultz, 1990):

- If all types of pipe are available, the use of corrugated plastic pipes has distinct advantages
- If pipes are not locally available, local manufacture of concrete pipes is the most straightforward and the easiest to implement. It requires less skill than manufacturing other types of pipe, and is already economical on a small scale. Plastic pipes occupy an intermediate position: local manufacturing from imported raw material is indeed possible for reasonably large quantities.
- Plastic pipes are particularly suited for machine installation. They have the advantage that their performance is the least affected by poor installation practice.
- The manufacturing cost of small diameter pipe (i.e. < 100 mm) is usually of the same order for clay tiles, concrete tiles and plastic. For large diameter pipes, however, concrete is usually the cheapest and plastic the most expensive.

Types of drainage tiles/pipes for field drain

- Perforated plastic (PVC or HDPE);
- Concrete;
- Clay

Plastic drain pipes (either PVC or HDPE) are currently the most common and the easiest and most secure types of pipes for machine installation.

The plastic material is inert and is not affected by soil chemicals.

Concrete pipes can be made relatively simply, onsite. The installation of drains with concrete pipes is more cumbersome and much more susceptible to misalignment than the installation of plastic pipes.

Under unstable soil conditions it is almost impossible to properly install concrete pipes. Some soil chemicals attack concrete pipes. When making cost comparisons the risks of breakage and possible high transport costs have to be taken into account.

The quality of clay pipes or tiles is comparable to concrete pipes. They are somewhat lighter than concrete pipe and are chemically resistant to soil chemicals.

6.2. CHARACTERISTICS OF TILES

The good characteristics of tiles are

1. Weather proof and deterioration free
2. Good strength for different load conditions
3. High density
4. Low water absorption
5. Resistance to freezing and dissolving;
6. Freedom from defects
7. Uniform thickness and circular shape.

6.3. CLAY TILES

Clay tile pipe sections are abutted against each other and water enters through the joints. The porous type usually has butt joints, but it may also have flanges or collars or bell joints. The latter type of tile is more expensive, and the extra cost is only justified in very soft soils. Good quality pipes are adequately baked and are free from cracks and blisters. Clay tile with cracks or other visible shortcomings and badly formed pipes should not be used. Clay drain tile is classified as either common or vitrified. The common tile, the one most used is made from common brick clay and is burned in the same manner as brick vitrified tile is made of ground shale or of a high grade clay, frequency mixed with common clay. It is other glazed is stronger and less porous than common clay tile. Ordinarily however this is the form of tile is not used for agriculture drainage because of its high cost. Clay pipes are cylindrical in form and about 0.30 m in length. Ends are either straight or sometimes have "collars" (effective but expensive). Clay tiles should be well burnt and without any defects. Good clay tiles have distinct ring when tapped with a metal object. Clay tiles have the advantage of being cheaper than concrete tiles and also can be used in acid and alkaline soils. Standard drainpipe sizes are 50, 65, 75, 80, 100, 130, 160, and 200 mm inside diameter. The wall thickness varies between 12 and 24 mm. However, they are difficult to transport without breakages and will not be able to withstand as much load as concrete tiles can take. Clay tile is very durable and highly resistant to weathering and deterioration in aggressive soil conditions e.g. in soils containing sulphates and corrosive chemicals. It can be used in almost all circumstances. Clay tile is lighter than concrete and has excellent bearing strength. It is however, fragile and must be handled with care. Clay tiles require a good deal of manual handling, although manufacturers have improved this by various methods of bulk handling.

Base material

Clay pipes are produced from clay through vitrification in an oven at a high temperature (>900°C). For the production of high quality pipes the clay should have no or limited content of montmorrilonite.

Fabrication of clay tiles

Clay pipes have to be produced in a specialized plant.

The mixing and wetting of the clay, the pressing through the mould, the cutting, the storing and the baking in the oven are all specialized jobs requiring specialized equipment. The baking process is highly sensitive, since the vitrification of the clay minerals only takes places at specific high temperatures that are to be maintained during a number of days. It also demands a considerable amount of energy. Manufacturing of clay tiles requires a great deal of skill and a well-equipped plant. The major quality features are straight joints, absence of cracks and homogeneity of the raw material (well-mixed clay). The maximum water absorption rate after being immersed in water for 24 hours should be less than 15 per cent of the weight of the tile. The weight of 1000 tiles should exceed certain minimum values, e.g. 1400 kg for 60 mm diameter pipe and 2000 kg for 80 mm diameter pipe. In some areas, clay and concrete tiles are still laid manually in a hand dug or mechanically excavated trench. These pipes may be covered with bulky materials or with 'envelopes' in strip form. Clay tiles should be installed in such a way that a perfect alignment between individual pipes is obtained. The maximum gap between individual pipes may not exceed 3 mm, except for sand where it should be not more than 2*d*85, i.e. the particle size for which 85 per cent of the soil particles on dry weight basis have a smaller diameter.

6.4. CONCRETE TILES

Concrete pipes for field drains generally come in a cylindrical form in lengths of about 0.30 m. It is used if clay tile is not available or if greater diameters must be applied. Concrete pipes are used mostly in medium to large sizes, with inside diameters of 60 to 100 mm are common inter diameters. Tile over 300 mm inside diameter is usually reinforced. The manufacture of concrete tiles is much simpler than that of clay tiles. Pipes should be well formed, finished, free from cracks and chips, and properly cured.

Base material

Concrete pipes are produced from cement and sand. If concrete pipes are used in acid or sulphuric soils, sulphate resistant cement must be used, in which case the density of the concrete has to be high so that it absorbs little or no water.

Suitability of Concrete pipes

Concrete pipes should be used only when soil and groundwater analyses have established that conditions are suitable for their use. Pipes made with ordinary cement are liable to deteriorate in acidic and high sulphate soils, and by water carrying certain alkali salts or other chemicals. Concrete pipes should not be used at locations where industrial waste or house refuse has been collected. Special high sulphate-resistant cements and high density concrete should be used to resist chemical attack (e.g. by acids or sulphates). Concrete pipes may disintegrate slowly from weathering, and are subject to erosion from fast flowing water carrying abrasive material.

Fabrication of concrete drain pipes

Concrete tile can also be locally made using simple moulds made of mild steel sheets. Concrete in 1:2:4 proportions should be used and the tiles should be properly cured. The pipe production in a factory usually has a better and more homogeneous quality. It is up to the manufacturer to prove that the quality standards (often national standards) are met and that they are checked by an independent entity. Moulds, preferably vibration tables, weighing units, hoppers and curing basins are required for onsite production of pipes. Quality production requires considerable supervision and quality control needs constant attention.

6.5. PLASTIC DRAIN PIPES

Plastic pipes can be made of polyvinyl chloride (PVC), high density polyethylene (HDPE), and to a minor extent, polypropylene (PP). The major differences between the base materials are:

- All plastic materials are sensitive to ultra violet (UV) radiation and should not be exposed to direct sunlight therefore.
- HDPE is sensitive to deformation at high temperatures
- PVC becomes brittle at low temperatures (near freezing point), HDPE also becomes brittle, but at somewhat lower temperatures and may thus be installed at lower temperatures. Once installed neither PVC nor HDPE are affected by low temperatures. Smooth plastic pipes were made of rigid polyvinyl chloride (PVC) and were provided with longitudinal slits to permit water entry. Smooth plastic pipes have never found a widespread use because they were rapidly superseded by corrugated pipes.
- HDPE requires approximately 20% more base material in weight per meter than PVC to attain the same strength.

Types of plastic pipes

Rigid smooth pipes : Rigid pipes are less easy to install and are considerably more expensive than corrugated pipes.

Corrugated pipes : To achieve the same strength and resistance to outside pressure a rigid pipe requires 3 x the amount of base material and, consequently, is about 3 times more expensive. The corrugated shape of the wall makes the pipe not only flexible, but also more resistant to compression than the smooth pipe, for the same quantity of raw material. In addition, corrugated pipe has facilitated the development of trenchless installation techniques.

Advantages of Corrugated pipes

- Light weight makes handling easier, even for great lengths.
- Long, continuous length eases handling, gives less alignment problems, and reduces stagnation of pipe supply resulting in a high installation rate for drainage machines.
- Flexibility and coilability facilitate handling, transportation and installation.
- Greater and more uniformly distributed perforation area, facilitating access of water.
- Easy wrapping with envelope materials.
- Safer implementation without too wide joints or misalignment.
- Less labour intensive and consequently lower labour cost for manufacture, handling, transportation and installation.
- Inert to all common soil chemicals.

Disadvantages of Corrugated pipes

- Vulnerability to deterioration from UV-radiation when exposed to sunlight for long periods, especially if made of PVC.
- Increased brittleness at low temperatures.
- Increased deflection risk at high temperatures and excessive stretch during installation.
- Lower deflection resistance under permanent load.
- Risk of collapse under sudden load, e.g. by trench wall caving or stones.
- Smaller transport capacity for the same inner diameter because of corrugation roughness.
- Not fire resistant.
- Not easy to relocate in the field with a tile probe without damaging the pipe.

Available sizes

Corrugated plastic pipes are available in diameters ranging from 40 to 600 mm. The smaller diameter pipes can be produced in coils in lengths of up to 300 m weighing between 30 and 50 kg. The larger the diameter, the smaller the length that can be coiled, e.g., 200 mm PVC pipes can be coiled up to a length of 50 m and the largest diameter pipes are seldom delivered on coils but as straight lengths of 6 or 9 m.

Fabrication of Corrugated Plastic Pipes

The production of corrugated plastic pipes is a specialised job. It requires expensive production lines consisting of: extruders, corrugators, perforators and coilers. The production capacity of a modern line is about 500 kg/per hour, for an Ø

100 mm pipe it amounts to approximately 1000 m/hour. The extruder can also be used for the production of rigid pipes. The corrugator and the perforator are parts of the production line that are specifically made the production of corrugated drain pipes. The corrugator gives the pipe its form; the form of the corrugation determines the strength of the pipe. Sophisticated corrugation forms can give the pipe the required strength at minimal wall thickness. The smaller the required wall thickness, the less the base material needed and thus the cheaper the pipe.

The perforator has to provide perfect holes in the pipe for water entry at the indicated place (in the valley of the corrugation) so that the perforation does not influence the strength of the pipe. The perforator should make holes and not slits or cuts. This means that the cut-out material has to be removed from the pipe. Slits and cuts tend to bend back and close the perforation over time. There are production lines for roughly Ø 40 - 100 mm and for Ø 100 - 200 mm. For each diameter to be produced on a production line, special tooling is required on the extruder, corrugator and perforator.

Double walled pipes (corrugated outer pipe and smooth inner pipe).

The slightly rougher wall of the smaller diameter corrugated plastic pipes will require marginally larger diameters. The extra costs thereof are negligible compared to the extra cost of either rigid or double walled pipes. The double-walled pipes were originally developed specifically for irrigation and can withstand higher than necessary internal pressures.

Physical properties of PVC and PE pipes

- The lower stiffness of PE means that pipes may be easily deformed under load, especially at temperatures approaching 40ºC, and if they are subjected to longitudinal stress.
- PVC pipes are more susceptible to UV-radiation and become brittle at exposure; storage of unprotected pipes in the open should therefore be avoided.
- PVC pipes are more brittle at low temperatures than PE pipes; PVC pipes should not be installed at outside temperatures below 3ºC because the risk that cracks will be formed is too high.
- PVC softens at 80ºC and drainpipes will deform when exposed to such temperature.
- Especially in arid and semiarid areas, special care shall be taken to prevent such storage conditions.
- PVC has some environmental disadvantages: it forms hydrochloric acid when burnt.

6.6. PIPE ACCESSORIES

Subsurface drainage systems require accessories and special structures such as pipe fittings (couplers, reducers, junctions, end caps), gravity or pumped outlets, junction boxes, inspection chambers (manholes), drain bridges, non-perforated rigid pipes, blind inlets, surface inlets, controlled drainage or subirrigation facilities, and cleaning provisions. Some fittings are made by pipe manufacturers, others are manufactured by specialized companies, and others are fabricated on the spot.

(a) End caps

End caps prevent the entrance of soil at the upstream drain-end opening. They can be made of the corresponding pipe material but any other durable flat material can be used for this purpose as well.

(b) Couplers

Corrugated pipes generally have external 'snap-on' couplers to connect pipes of the same diameter. Alternatively, a piece of pipe of the same diameter that is split for easy placing around both pipe ends, and firmly wrapped with tape or wire to keep it in place during installation, can be used instead. Internal couplers can be used with the trenchless technique to prevent separation of connected pipes when passing through the pipe feeder device (Schwab and Fouss, 1999). Pipes can also be connected internally by making a slit in the end of the upstream pipe and forming a cone that is pushed into the end of the downstream pipe. Such connections are not very reliable and do impede the discharge of water and suspended solids.

(c) Pipe fittings

A wide range of pipe fittings, made of various raw materials, is commercially available for all kinds of pipes. Fittings for clay, concrete and corrugated plastic pipes are generally made by the various pipe manufacturers and therefore they are mostly not interchangeable Cross, T and Y-pieces connect laterals or collectors with collectors. Many fittings are fabricated with multiple sizes at the ends facilitating the connection of various sizes of collectors and laterals (Schwab and Fouss, 1999). The end sides of the fittings are cut off, or adapted by removing some parts in the field to attach to the appropriate diameter. Simple connections with elbows and T-pieces on top of the collector are nowadays used to connect laterals with collectors.

6.7. TESTING OF DRAINAGE TILES

Pipe drains have been in use for more than 300 years. Originally drains were formed with trenches filled with gravel and wooden splinters and then topped with a layer soil. During the last 100 years clayware called drain tiles have been in use. Then cement concrete pipes came to be used. Since the sixties, plastic pipes have entered the usage in a big way. All Standards-Institutions, such as Bureau of Indian Standards, British Standard Institution, American Society of Testing Materials and German Standards (DIN) have all brought out specifications on materials to be used for buried drains.

However, the essential and desirable characteristics of drain pipes include the following:

(i) Good resistance to deterioration in the soil due to weathering,

(ii) Adequate physical strength to support the soil overburden, as well as live loads expected due to men, animals and machines, and

(iii) Uniformity of wall thickness and shape to ensure free and easy flow of drainage water.

Clay and concrete pipes come under the rigid variety, whereas plastic pipes (PVC and High Density Poly Ethylene) fall under the flexible variety.

Testing Parameters for Drain pipes:

For drainpipes the inside and outside diameter are specified with their tolerances. Moreover, the following parameters are usually included in standards:

(a) Clay and concrete pipes

- Ovality and curvature;
- Verticality of the end planes;
- Resistance to weathering and deterioration in soil;
- Resistance to freezing and thawing cycles;
- Density;
- Water absorption; and
- Crushing strength.

(b) Concrete pipes

In addition to the above:

- Sulphate resistance; and
- Acid resistance.

(c) Plastic pipes

- Stiffness and elongation resistance;
- Impact strength and brittleness;
- Flexibility and coilability;
- Perforations and hydraulic properties; and
- Handling and installation instructions.

The substitution of clay and concrete pipes by corrugated plastic pipes made standards for clay and concrete pipes less important although they are still useful in countries where clay and concrete pipes are still installed, including larger diameter collector drains.

Virgin material

Material in a form such as granules or powder that have not been subjected to use or processing other than that required for its manufacture and to which no reprocessable or recyclable materials have been added.

Recyclable material

Material comprising either one of the following forms :

(a) Material from used PVC-U pipes or fittings which have been cleaned and crushed or ground.

(b) Material from used PVC-U products other than pipes or fittings which have been cleaned and crushed or ground.

Nominal diameter (DN)

A numerical designation of diameter which is common to all components in a piping system. It is a convenient round number for reference purposes approximate to the manufacturing diameter, expressed in mm. For this system standard, it is based on the outside diameter of the corrugated pipes. For Scandinavia, nominal diameters are based on inside diameter.

Materials

Minimum PVC content

When tested in accordance with CEN/TC 155 WI 137, the content of PVC shall be at least 80 per cent by mass for pipes and 88 per cent by mass for fittings. In case of use of virgin and own reprocessable material, the minimum PVC content can be calculated.

Reference Conditions for Testing

The mechanical and physical properties specified in all Parts of this standard shall, unless otherwise specified, be determined at 23 ± 2°C.

Pipes without Envelope

Nominal diameter (*DN*)

Numerical designation of the outside diameter (*Do*) of the pipe declared by the manufacturer. For Scandinavia, nominal diameter is based on internal diameter (*Di*).

Mean outside diameter

The measured length of the outer circumference of the pipe, divided by π (= 3.142) and rounded to the next higher 0.1 mm.

Total length

The distance between two planes normal to the pipe axis and passing through the extreme end points of the pipes measured along the axis of the pipe.

Nominal length

Numerical designation of a pipe length declared by the manufacturer which is equal to the pipe's total length in metres stated as a whole number.

Ring stiffness

The value of initial resistance to radial deflection under external load obtained by testing in accordance with ISO 9969.

Creep ratio

A physical characteristic of the pipe obtained by testing in accordance with ISO 9967. It is a measure of the long-term resistance to radial deflection under external load.

Test method

On each test piece, mark four generating lines with an angle of approximately 45° between them and in a plane square to the pipe axis. Using a slide calliper, measure the four corresponding diameters and record the four individual measurements. Calculate the difference between the highest value and the lowest value and relate the difference to the nominal value as specified in above.

Minimum inside diameters

When measured to an accuracy of 0.1 mm or 0.05% whichever is the greater value, the average of the measured mean inside diameters shall not be less than the minimum Di given in Table for the relevant nominal diameter, DN. An internal micrometer or a plug gauge with an accuracy of 0.1 mm shall be used for the measurement of the inside diameter up to 180 mm. Above D_i 180 mm, any suitable measurement device may be used.

Length of test pieces

The length L, in metres, of the test pieces shall be as follows

$L = 0.2 \pm 5$ % for pipes with $DN \leq 200$;

$L = 0.4 \pm 5$ % for pipes with $DN > 200$.

Test method

On each test piece, mark four generating lines with an angle of approximately 45° between them and in a plane square to the pipe axis. Using a slide calliper, measure the four corresponding diameters and record the four individual measurements. Calculate the difference between the highest value and the lowest value and relate the difference to the nominal value as specified in above.

6.8. MECHANICAL CHARACTERISTICS

Impact resistance

When tested, the following requirements shall be conformed to as applicable:

(a) If $50 \leq \text{DN} \leq 200$, the pipes shall comply with the four following requirements: $(X_f + X_p)/2 = \text{H50 } 0.9$ m

$H_{50sl} \geq 0.6$ m, $H_{50pl} \geq 0.6$ m, $H_{i\ min} \geq 0.4$ m

(b) If DN > 200, the pipe shall conform with the four following requirements : $(X_f + X_p)/2 = H_{50} \geq 1.2$ m

$H_{50sl} \geq 0.9$ m, $H_{50pl} \geq 0.9$ m, $H_{i\ min} \geq 0.6$ m where: X_f is the average of the dropping heights when failure occurred;

X_p is the average of the dropping heights when test pieces passed;

H_{50sl} designates the seam lines H_{50};

H_{50pl} designates the perforation lines H_{50};

H_i min designates the minimum fall height without failure of the test.

Extensibility

This characteristic is not applicable for DN > 200.

When tested, no test piece shall have an elongation greater than 55 mm.

If the first test piece has an elongation less than 45 mm, the result is considered to be satisfactory.

If the first test piece has an elongation between 45 mm and 55 mm, the average of the elongations of this test piece with the two additional ones shall be less than 50 mm.

Brittle fracture test (rapid tensile test)

When tested, disregarding the first failure occurring within one nominal diameter, of the pipe being tested, from the anchoring devices, the result from three test pieces shall not include more than one failure. If one failure has occurred, the results from six further test pieces shall include no failures.

General

(a) The striker shall be type d90 with a mass of 1 kg.

(b) The circumferential orientation of the test piece in the V-block.

(c) Failure

A blow is considered as a failure if any of the following characteristics occurs

- The test piece breaks into two or more parts
- Fragmentation of the test piece occurs
- The test piece shows at least one crack joining continuously any couple of perforations
- A crack can be seen with the naked eye on the seam line and is longer than 5 mm.

Preliminary test procedure

Set the drop height of the striker at 0.4 m.

Main test

The main test is also divided into two series. Here, each series includes 16 test pieces. In series one, ensure that each test piece is hit by the striker on a perforation line selected at random. In series two, ensure that the test pieces are hit by the striker alternately on seam line one and on seam line two.

Record the dropping height values for the test pieces and note whether or not the test piece failed.

Calculate the H_{50} failure level using the following equation :

$$H_{50} = (X_p + X_f)/2$$

where

$X_5 =$ the average of the dropping heights when failure occurred;

$X_p =$ the average of the dropping heights when the test pieces passed.

Calculate two values of H_{50}, designated H_{50pl}, and H_{50sl} as follows :

H_{50pl} is the value derived from the 16 blows on the perforation lines;

H_{50sl} is the value derived from the 16 blows on the seam lines.

A blow is considered as a failure if :

- the test piece breaks into two or more parts ;
- fragmentation of the test piece occurs (detail A) ;
- the test piece shows at least one crack joining continuously any couple of perforations
- a crack can be seen with the naked eye on the seam line and is longer than 5 mm.

6.9. TESTING THE STRENGTH OF CONCRETE OR CLAY TILES

(a) The three edge bearing test:

- The equipment for these tiles is consisting of two wooden blocks on which the tiles are placed.
- A point bad which can be increased is applied on top of the tile. This load is applied either by mechanical Jacker a hydraulic system.
- The load is gradually increased and the load at which the tiles break is noted.

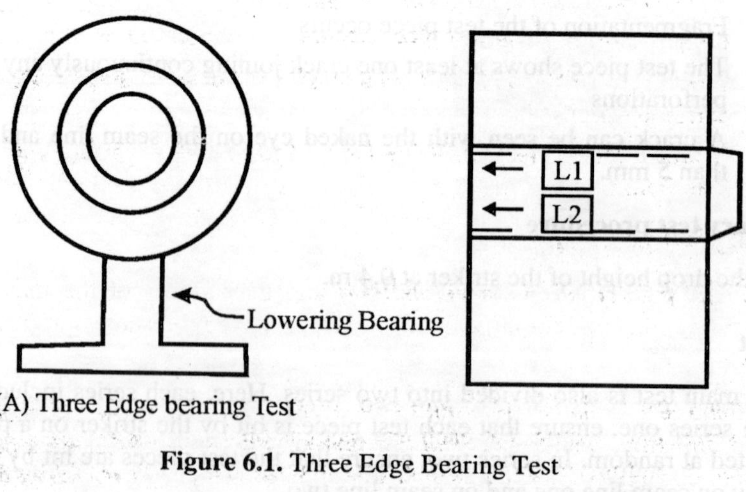

Figure 6.1. Three Edge Bearing Test

(b) The sand bearing test

- In the sand bearing test the procedure is the same as the three edge bearing test but the arrangement is different.
- In this method the bottom of the tile is supported by a bed or of sand. This represents the conditions found under normal installation procedure more closely than the strength found by the three edge method.
- The testing of plastic pipes used for tiles is done differently plastic do not get crushed like clay or concrete tiles but instead get deformed under load. Deformation is a characteristic that is measured in pipes.

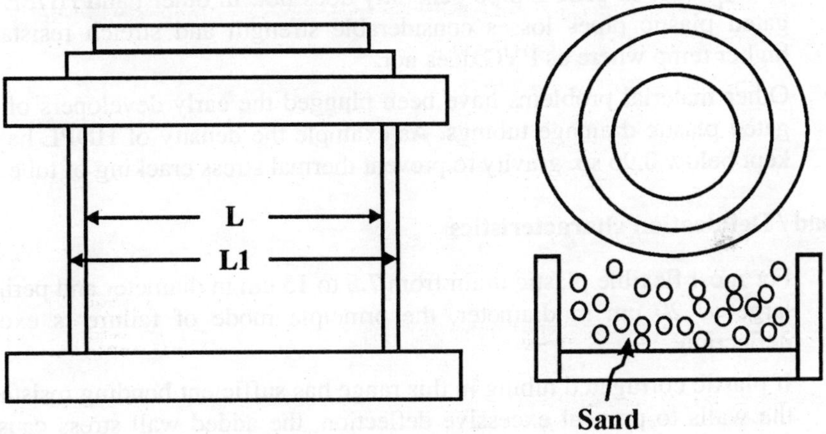

Figure 6.2. Sand Bearing Test

(c) Sand box test

- Specification for corrugated polythelyne drainage tubings is based upon a sand box loading test.
- Dry off sand is used to surrounding the teats specimen. The minimum dimension for test box are 3D square (where D is the tubing inside diameter) by 3.75D deep.
- The top specification calls for minimum crushing strength of 2460 g/cm^2 (=35.0 lb/in^2) unit loading under the steel loading plate.
- The sand box test method workable for the 10 cm inside diameter Tubings and filled valuable need during the easy development and use of the corrugated plastic tubings.
- The size requirement for the box and the validity of the minimum crush strength value for large diameter Drains have represented problems.

6.10 POSSIBLE FUTURE TEST METHODS

The principle features of the proposed specified for material and testing methods of corrugated plastic tubes are given below.

(a) Types of plastic

- The most common plastic used for corrugated tubings are high density polyethylene (HDPE) and PVC because of the supply demand relationship. PVC plastic material is considerably stronger than HDPE. HDPE has been used successfully in small diameter Corrugated plastic drains (e.g. 7.5 to 15 cm inside diameter), but with the higher strength requirement for the large tubes, the use of PVC or other material may be considered.
- The strength of thermo plastic materials handling problems, especially during installation. A PVC plastic pipe tends to become brittle at low temp (0°C) where as HDPE pipe generally does not. In other hand HDPE corrugated plastic pipes losses considerable strength and stretch resistance at higher temp where as PVC does not.
- Other material problems have been plugged the early developers of corrugated plastic drainage tubings. As example the density of HDPE has to be kept below 0.96 sp. gravity to prevent thermal stress cracking of tube wells.

(b) Load / Defalcation characteristics

- For most flexible plastic drain from 7.5 to 15 cm in diameter and perhaps as large as 20 cm in diameter, the principle mode of failure is excessive deflection.
- If plastic corrugated tubing in this range has sufficient bending resistance in the walls to prevent excessive deflection, the added wall stress caused by ring compression under soil loading device to measure deflection resistance of plastic tubings seen justified.
- External loading properties of plastic by parallel plate bedding, one main exception proposed for this test procedure is to reduce the rate of advance of the parallel plates towards of each other to deflect the tube specimen below the 1.27 cm/min rate specified as Diameter cm/min , D= diameter Of the tube in cm is recommended.
- The parallel plate strength of the specimen is expressed in terms of

$\omega/\Delta y$

at a specified per cent deflections $100\Delta y/D$,

where ω = parallel plate load, g/cm of tube length, Δy = vertical deflection.

(c) Impact resistance

- In addition to deflection resistance test to detect excessively brittle or poor quality plastic drain tubing especially that with nearly invisible stress cracks or faults. To improve the effectiveness of the impact test, all tube specimens should be preconditioned and the test conducted at low (0°C) temp.
- For corrugated plastic tubing, a type B tap way proposed which has 5 cm radius of curvature on the striking surface rather than the pointed type A tap commonly used for testing smooth –wall plastic pipe.

- An attendance proposal is used to striker namely, a 1.25 am diameter Steel rod impact longitudinally along, the top of the tube specimen. So that several corrugations are stack at once.
- A tentative minimum impact resistance of 1.5 kg.m was specified which is typical for was specified which is typical for PVC corrugated plastic tubings fabrication.

6.11 ENVELOPE FOR SUBSURFACE DRAINAGE SYSTEM

The material placed around a subsurface drain, to protect the drain from sedimentation and improve its hydraulic performance, should be referred to as a drain envelope. Porous material placed around a subsurface drain, to protect the drain from sedimentation and improve its hydraulic performance, should be referred to as a drain envelope. It is worthwhile to distinguish between the definition and function of an envelope and that of a filter. Functions of envelope are: Filter function: to prevent or restrict the soil particles from entering the pipe where they may settle and eventually clog the pipe; Hydraulic function: to constitute a medium of good permeability around the pipe and thus reduce entrance resistance; Bedding function: to provide all-round support to the pipe in order to prevent damage due to the soil load. Note that large-diameter plastic pipe is embedded in gravel especially for this purpose. It prevents the entry of soil particles into the pipe, although a limited flow of clay particles will do little harm, because they mainly leave the pipe in suspension. Envelopes have the task to improve the permeability around the pipe, and act as permeable constraints to impede entry of damaging quantities of soil particles and soil aggregates into drainpipes. Yet the majority of small particles of soil material and organic matter, suspended in water moving toward a drain, will actually pass through a properly selected and installed drain envelope without causing clogging. The relatively coarse envelope material placed around the drain should stabilize the soil mechanically and hydraulically, but should not act as a filter. Envelope materials used to protect subsurface drains have included almost all permeable porous materials that are economically available in large quantities. Based on the composition of the substances used, they can be divided into three general categories: mineral, organic, and synthetic envelopes.

Selection of Envelope

Qualitative guidelines for designing drain envelopes mainly consider soil texture. Straightforward rules can be given for fine- and coarse-textured soils. For soils in the intermediate texture classes, there is considerable uncertainty. Fine-textured soils with a clay content of more than about 0.25 to 0.30 are characterized by a high structural stability, even if being worked under wet conditions. Thus, with trencher-installed pipe drains, no problems are to be expected and an envelope is not required. With trenchless drainage, however, one could easily work below the critical depth, especially in wet conditions, resulting in a high entrance resistance. An envelope is not likely to be of any help. Clogging of the pipe is not to be expected. Coarse-textured soils free of silt and clay, on the other hand, are permanently unstable, even if undisturbed. Thus, soil particles are likely to wash into the pipe, both from the trench backfill and from the undisturbed soil below the pipe. There is a need for a permanent envelope, completely surrounding the pipe, only as an effective filter, because there is no high entrance resistance. A thin geotextile envelope is probably the best solution here. Soils of intermediate texture are less simple. In the finer-textured soils of this category (clay contents less than 0.25 to 0.30, but more than say 0.10 to 0.15), the installation is done under dry conditions and, in irrigated land, provided that the trench backfill was properly compacted. In those cases, even without an envelope, no problems will arise. If, however, the pipes were installed under wet conditions, both drain sedimentation and a high entrance resistance could follow. Hence an envelope would be needed. Most likely, only the trench backfill will create problems, because the undisturbed soil remains stable enough. At the coarse-textured side of the intermediate soils (soils with a clay content below 5% and a high silt content), the trench backfill is likely to be as unstable as the undisturbed soil below the pipe. In addition, the trench backfill may become poorly permeable through a re-arrangement of the soil particles. Therefore, an envelope which completely surrounds the drain, fulfilling both filter and hydraulic functions, is always needed in these soils.

6.12 ENVELOPE MATERIAL

(a) Granular envelopes

Mineral envelopes mainly consist of coarse sand, fine gravel and crushed stone, which are placed under and around the drainpipe during installation. If well designed and installed, mineral granular envelopes are quite reliable because they are voluminous and can store comparatively large quantities of soil material without noticeable malfunctioning. As such, they have provided satisfactory long-term service under most circumstances. Traditionally, pit run naturally graded coarse sand or fine gravel containing a minimum of fines is the most common and widely used drain envelope material. Such material can be as permanent as the soil itself. Properly designed graded gravel envelopes fulfill all the mechanical and hydraulic functions of a drain envelope and are the ideal envelope from a physical standpoint. Graded gravel should be a homogeneous, well-graded mixture of clean sand and gravel free from silt, clay, and organic matter, which could adversely affect its permeability. The use of limestone particles must be avoided, because a high per centage of lime in gravel envelopes is a source of incrustation. In addition, the gradation of a gravel envelope should be made in accordance to prescribed parameters.

(b) Organic envelopes

Organic materials, many of which are by-products of agricultural production, have successfully been applied as drain envelopes. They are voluminous, so they can be used in cases where both particle retention and hydraulic function are important. Organic materials may be applied directly on the drainpipe in the trench as loose blinding material, or may be prewrapped around the drainpipe as Prewrapped Loose Materials (PLMs). An intermediate type of application has been in strip-form, applied on top of the drainpipe. This type of application is now obsolete. Organic envelope materials include chaff, cereal straw, flax straw, rice straw, cedar leaf, bamboo, corncobs, wood chips, reeds, heather bushes, chopped flax, flax stems, grass sod, peat litter and coconut fibre. The flax straw was applied originally as a cover strip and later as prewrapped envelope. The coarseness of the flax envelope did however not always guarantee the particle retention function. On a much smaller scale, other organic envelopes have been applied. These materials were not always available in the required quantities and their handling was often laborious. The use of straw was not successful because it usually decomposed into a low-permeability layer around the pipe. At the end of the 1960s, coconut fibre was introduced, being relatively cheap, it soon dominated the market because high quality peat litter became scarce and expensive and because the flax industry declined. Moreover, the finer coconut fibre was considered a more appropriate envelope material than the coarser- structured flax straw. The envelopes were usually fully decomposed after two to five years, particularly if the pH of the soil exceeded the value 6. The organic envelopes may affect chemical reactions in the abutting soil. This process may result in biochemical clogging of the drain. If iron ochre clogging of drains is likely, reluctance with the application of organic envelopes is justified. Even organic matter that is accidentally mixed with trench backfill material may severely enhance the risk of ochre clogging of the drain

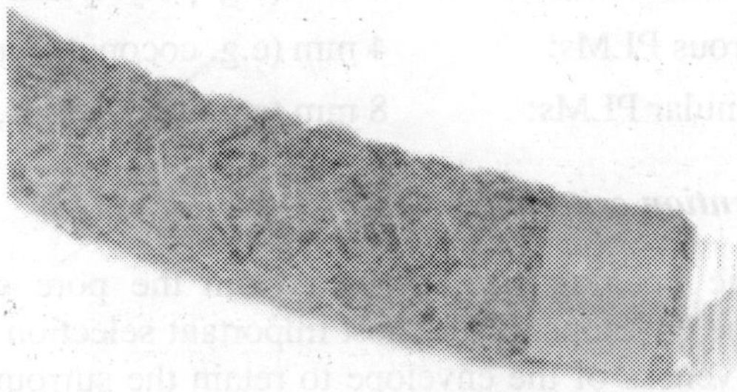

Figure 6.3. Coconut Fiber Envelope

(c) Synthetic envelopes

Many of the drawbacks of gravel envelopes can be overcome with the use of synthetic envelopes. A synthetic envelope is a permeable structure consisting of loose, randomly oriented yarns, fibres, filaments, grains, granules or beads, surrounding a corrugated drainpipe, and retained in place by appropriate netting and/or twines. Synthetic envelopes are usually wrapped around the corrugated plastic drainpipes by specialized companies and occasionally in pipe manufacturing plants. The finished product must be sufficiently strong to resist handling and installation without damage. Synthetic envelope include various polymeric materials. Fibres may be made of polyamide (PA), polyester (PETP1), polyethylene (PE), and polypropylene (PP). Loose polystyrene (PS) beads can be wrapped around drains in perforated foil or in string netting.

Synthetic envelope materials deteriorate when exposed to solar (UV) radiation. The speed of the deterioration will be double in semi-arid and arid regions where the average annual radiation is twice that in temperate regions. However, once installed, synthetic envelopes, manufactured from suitable raw material are not subject to decomposition. These materials are therefore, reliable and affordable substitutes for conventional gravel and organic envelopes.

6.13. DESIGN CRITERIA OF ENVELOPE

Thickness of envelope

The thickness of envelopes serves as a reference for uniformity and conformity. In addition, envelope thickness is found a factor of importance in theoretical analyses as it influences the soil retention capacity, the entrance resistance of drainpipes and the exit gradient at the soil- envelope interface. The main task of an envelope is soil particle retention. In this respect, design criteria for envelope thickness are irrelevant. Thicker envelopes, however, may have higher porosities, which explain their popularity when chemical clogging is anticipated. Therefore, in the envelope selection procedure, envelope thickness is an important parameter, and often significant in terms of safety.

According to the provisional EN-standard (CEN/TC155/WG18, 1994), the following minimum thicknesses are required:

- Synthetic, fibrous PLMs: 3 mm (e.g. PP fibres).
- Synthetic, granular PLMs: 8 mm (e.g. polystyrene beads).
- Organic, fibrous PLMs: 4 mm (e.g. coconut fibres).
- Organic, granular PLMs: 8 mm (e.g. wood chips, sawdust).

Opening size and retention criterion

The characteristic opening size, derived from the pore size distribution or porometric curve of the envelope, is the most important selection criterion because it determines the effectiveness of the envelope to retain the surrounding soil material. The retention of soil particles is normally not a problem since very fine fabrics are available. Laboratory research as well as practical experience, however, have revealed that fine envelopes are vulnerable to mineral blocking and clogging. Blocking of an envelope is a decrease of the number of active openings in an envelope that occurs when it is brought in contact with a soil. Clogging, on the other hand, is a decrease with time of the number of active openings in an envelope due to gradual accumulation of particles inside and on its surface, e.g. by particles suspended in turbid water. Therefore, specifications for envelopes should cover both soil retention criteria and criteria to prevent clogging and blocking of the envelope. Intensive research has resulted in criteria for soil particle retention and in recommendations with respect to the problems of blocking and clogging.

Hydraulic conductivity

The hydraulic conductivity of envelopes should be greater than that of the soil in order to reduce the entrance resistance of drainpipes, so that no hydraulic pressure will develop outside the envelope. The hydraulic conductivity, perpendicularly to or in the plane of envelope, can hardly be a problem because envelopes are much more permeable than the adjacent soil that they have to retain. Even under load, the hydraulic conductivity of compressible envelopes will meet the conductivity requirements.

6.14. DESIGN OF DRAINAGE ENVELOPE

The standard procedure for the design of gravel envelopes is to match the particle-size distribution of the soil with the particle-size distribution of the gravel. Several sets of design criteria to prevent base soil invasion into the envelope and the drain pipe have been developed. Filter specifications were first developed by Terzaghi for hydraulic structures. The base material is the soil found at the depth the tile is laid. Samples of the base material are obtained and a mechanical analysis made. Based on the design criteria developed by the U.S. Bureau of Reclamation and the U.S. Army Corps of Engineers, limits are established, which should be met by a filter material for a specific base material. Multiplying the 50 per cent grain size of the base soil material by 12 and 58 will give the limits, the 50 per cent size of the filter should fall within these limits. Multiplying the 15 per cent fine size of the base soil material by 12 and 40 provides the limits, the 15 per cent fine grain size of the filter should fall within these limits.

Nevertheless, although modern drainage machinery has facilities to install gravel automatically under and around the pipe, it remains a costly and difficult operation because of (Dierickx 1993):

- Uncertainties about gravel specifications and gravel shape (rounded or angular);
- Lack of uniform quality and gradation of gravel;
- Segregation during transport and installation;
- 'Flowability' problems in the supply chute of the drainage machine;
- Unequal distribution around the drain pipe;
- High demand on logistics

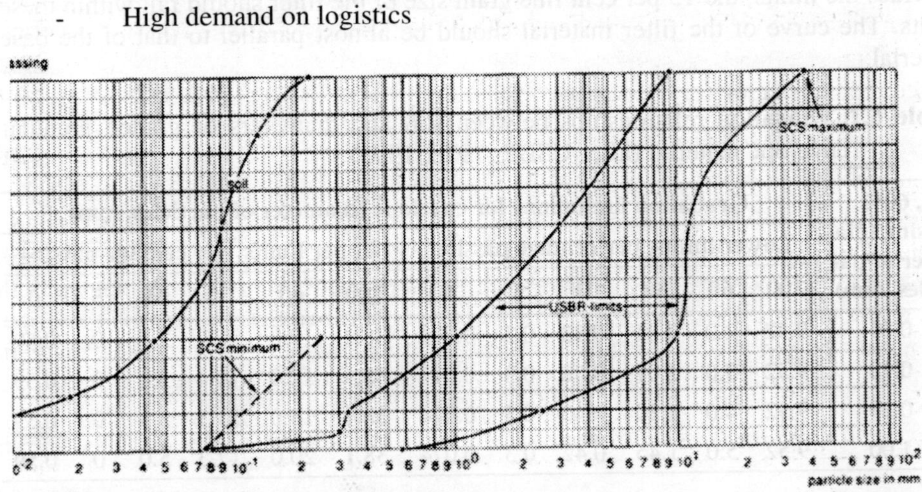

Figure 6.4. Example of the Design of a Grave Envelope

Design of Gravel Filter

Graded gravel should be a homogeneous, well-graded mixture of clean sand and gravel free from silt, clay, and organic matter, which could adversely affect its permeability. The use of limestone particles must be avoided, because a high per centage of lime in gravel envelopes is a source of incrustation. The filter is determined by the mechanical analysis and the gradation curve of the base material. To design the gravel envelope for a given soil make a mechanical particle size analysis of both the soil and the proposed gravel envelope; compare the two particle size distribution curves; and decide, by some design criterion, whether the proposed gravel envelope material is suitable.

Terzaghi's criteria

The first criteria, proposed by Terzaghi (US Army Corps of Engineers, 1941) for what he termed a 'filter', are:

The particle diameter of the 15 per cent size of the filter material (D_{15}) should be at least four times as large as the diameter of the 15 per cent size of the soil material (d_{15}):

$$D_{15} \geq 4\ d_{15}$$

This requirement would make the filter material roughly more than ten times as permeable as the soil. The 15 per cent size of the filter material (D_{15}) should not be more than four times as large as the 85 per cent size of the soil material (d_{85}):

$$D_{15} \leq 4\ d_{85}$$

This requirement would prevent the fine soil particles from washing through the filter material. Multiplying the 50 per cent grain size of the base soil material by 12 and 58 will give the limits; the 50 per cent size of the filter should fall within these limits. Multiplying the 15 per cent fine size of the base soil material by 12 and 40 provides the limits; the 15 per cent fine grain size of the filter should fall within these limits. The curve of the filter material should be almost parallel to that of the base material.

Table 6.1. Gradation relationships between soil and diameters of graded granular envelope material (after USBR, 1978, 1993)

Soil, 60% passing (diameter of particles, mm)	Gradation limitations for envelope (diameter of particles, mm)											
	Lower limits, per centage passing						Upper limits, per centage passing					
	100	60	30	10	5	0	100	60	30	10	5	0
0.02-0.05	9.52	2.0	0.81	0.33	0.3	0.074	38.1	10.0	8.7	2.5	0	0.59
0.05-0.10	9.52	3.0	1.07	0.38	0.3	0.074	38.1	12.0	10.4	3.0	0	0.59
0.10-0.25	9.52	4.0	1.30	0.40	0.3	0.074	38.1	15.0	13.1	3.8	0	0.59
0.25-1.00	9.52	5.0	1.45	0.42	0.3	0.074	38.1	20.0	17.3	5.0	0	0.59

Design of Organic Envelopes

The organic envelopes are the by-products of agricultural production. The criteria for the design of organic envelopes are based on the opening size of the envelope material, average diameter of soil particle, etc. The life of organic envelopes depends on micro-biological activity in the soil, which is a function of temperature, soil chemical properties, and the presence of oxygen. The rapid decay of coconut fiber envelopes has stimulated the search for affordable, synthetic alternatives.

Design of Synthetic Envelopes

The criteria for the design of synthetic envelopes are based on the opening size of the envelope material, hydraulic conductivity, average diameter of soil particle, coefficient of uniformity, hydraulic gradient. Fibers may be made of polyamide, polyester, polyethylene and polypropylene. Synthetic Pre-wrapped loose materials (PLM) envelopes are usually wrapped around the corrugated plastic drainpipes by specialized companies and occasionally in pipe manufacturing plants. The finished product must be sufficiently strong to resist handling and installation without damage. The envelope is limited to the pipe of 200 mm diameter or less. A geotextile is defined as 'a planar, permeable, polymeric (synthetic or natural) textile material, which may be woven, non-woven or knitted, used in contact with soils. Non-woven geotextiles are sheets or webs consisting of directionally or randomly oriented fibers, filaments, or other elements. These elements are bonded by mechanical, thermal and/or chemical means. Knitted geotextiles are manufactured by interlooping one or more yarns, fibers, filaments, or other elements.

According to the provisional EN-standard (CEN/TC155/WG18, 1994), the following minimum thicknesses are required:

Synthetic, fibrous PLMs:	3 mm (e.g. PP fibres).
Synthetic, granular PLMs:	8 mm (e.g. polystyrene beads).
Organic, fibrous PLMs:	4 mm (e.g. coconut fibres).
Organic, granular PLMs:	8 mm (e.g. wood chips, sawdust).

The hydraulic conductivity of envelopes should be greater than that of the soil in order to reduce the entrance resistance of drainpipes.

CHAPTER – 7

INSTALLATION AND MAINTENANCE OF DRAINAGE SYSTEMS

In planning and designing a subsurface drainage system, decisions are to be taken on the layout and arrangement of drain lines, selection of a suitable outlet, proper depth and spacing of laterals, gradients of collectors and conductors, length and diameter of drain pipes, material of the drains to lend adequate strength, and accessories such as inlets and outlet structures.

The main factors affecting in subsurface (underground) drainage design include : soil properties, rainfall and the resulting design discharge that can be expected, the site topography and land use (especially in deciding the layout), and the elevation of possible outlet locations (especially in deciding how deep the drain could be set, if pumping is to be avoided). If an impervious structure (rock, hard pan, tight clay etc.) exists at a finite depth from the ground surface, that also influences the design. If machinery is to be employed for trenching and pipe laying, the depth and side slopes of open ditches would be limited to that extent.

Based on a large number of systems that have already been built and operated, the following ranges are found to be typical: The depth of a drain, from ground surface, varies between 0.75 m and 2.5 m. The lower the permeability the smaller would be the depth. In heavy clay soils, the drain spacing have been between 5 and 15 m. With mole drains, 30 to 60 m spacing has been adopted. In very permeable soils, spacing can be upto 150 m. The gradients adopted for a drain could be as steep as 1 in 30 or as flat as 1 in 1000. The danger of clogging by sedimentation in flatter gradient should be guarded against. These ranges are to be taken as indicative only. Cases of 3 m drain depths, 200 m drain spacing, and 1 in 2000 gradient of the drain pipe also exist. For idealized conditions expressions have been obtained relating the depth of drain and its spacing to the flow rate through the drain along with other relevant parameters.

The length of the field drains is determined either by fixed (farm) boundaries or by a fixed length for the drain. It is often decided to place the field drains at right angles to the collectors. If so, it may happen that the field drains do not run parallel to the minor infrastructure (e.g. irrigation canals or farm roads). In such a case, it is better to install the field drains at such an angle to the collector that the number of crossings with the minor infrastructure is minimized.

The spacing of the collectors is often determined by the length of the field drains. The collector alignments are further fixed by the field boundaries. The length of a collector is restricted either by a field boundary or by the available slope. The available slope is fixed by the shallowest permissible drain depth, the maximum water level in the main drain, and the slope of the land surface.

The field drainage system is generally designed on a model basis for a sample area. This can be a single farm of less than one hectare or an area of more than hundreds of hectares. In such a sample area, the design variables (i.e. The soil and hydrological conditions and agricultural inputs) are considered to be uniform. This means that the design is only a guideline, which can be adjusted for each single farm to incorporate specific circumstances.

As in the case of surface ditch layout, we have a choice from among several alternative layout schemes, such as: (i) natural or random, (ii) herringbone, (iii) grid iron, and (iv) interceptor type of layout of subsurface pipe drains. Also, there could be a combination of pipe-segments and open ditches to complete a drainage system. Of the first three, the natural type follows a random layout in order to connect ground water pool locations as distributed in nature; the grid iron pattern is of a regular geometry with tertiary (equally spaced field drain or laterals), secondary (collector drains) and primary (main drain) segments. Whereas, the natural layout follows the direction of natural water movement the grid iron has both along and across segments. The intercepting layout is aligned normal to the seepage direction. Whereas the first three types are based on the shape of the contours of the area to be drained, the interceptor drains are dependent on the cross-sectional features of the drained area.

It is desirable that the layout of irrigation supply canal/channel systems and the layout of drains (especially the open ditch segment) be jointly decided upon to ensure maximum water control in the irrigated agricultural area.

7.1. LAYOUT OF SURFACE AND SUBSURFACE DRAINAGE SYSTEMS

The possibility of use one or more or a combination of these systems depends upon the site conditions. Field topography and outlet location/elevation are typically the major factors considered in planning drainage system layout, with topography greatly influencing what layout alternatives are possible. A number of methods may be used to create the map, including standard topography surveys, a GPS or a laser system. The topography map helps the designer assess overall grade and identify the high or low spots in a field that might pose challenges.

The different layouts for surface and subsurface drainage systems are as follow:

(a) Random drainage system

The random drainage system is used for draining isolated low laying water logged patches when the entire area does not need drainage. Drain lines are laid more or less at random to drain these wet areas. In most cases the drain main follows the largest natural depression in the field, and submains and laterals extend to the individual wet areas. If the individual wet areas are large, the arrangement of the submain and laterals for each wet area may utilize the gridiron or herringbone tile pattern to provide the required drainage.

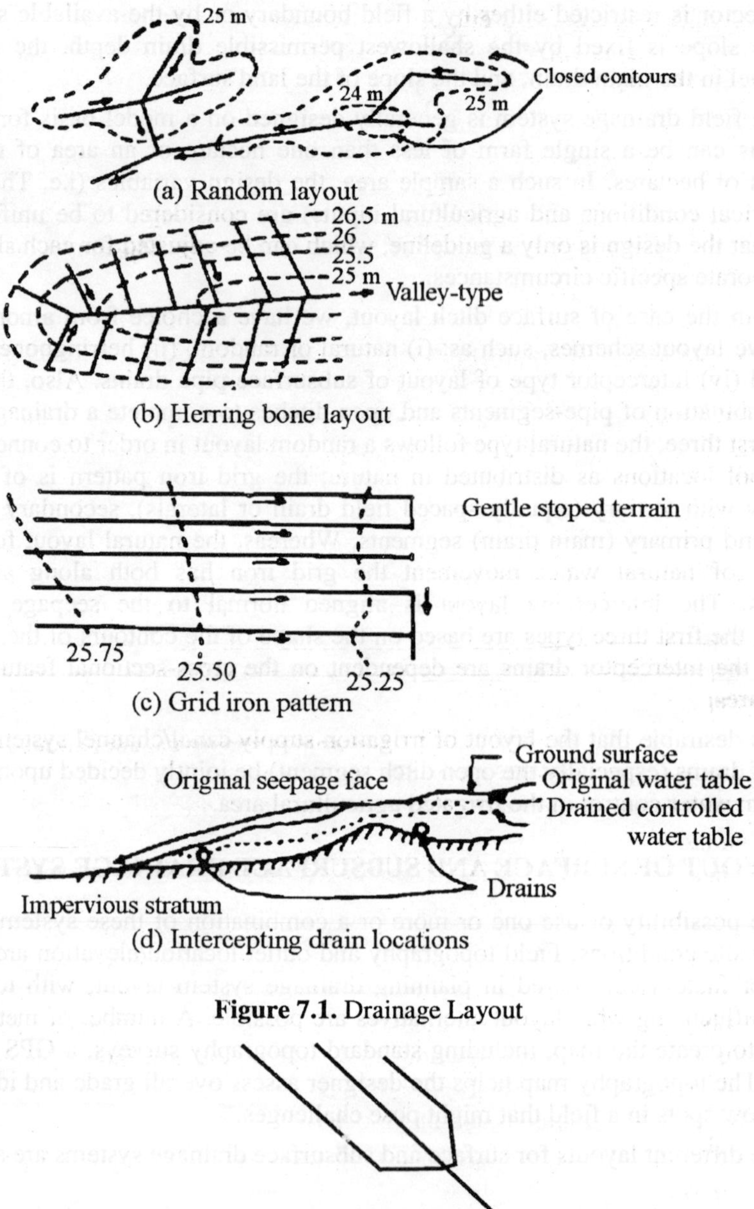

Figure 7.1. Drainage Layout

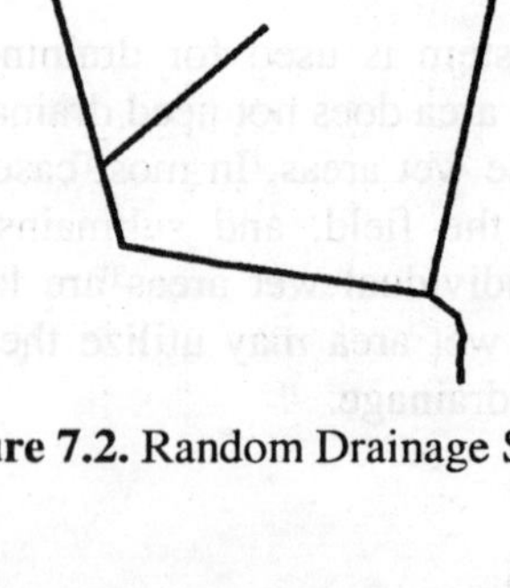

Figure 7.2. Random Drainage System

(b) Herringbone type drainage system

The herringbone type drainage system is used in areas where a mainline or submain could be laid both sides. The land in this system is well drained. This system is used where the main or sub main lies in a narrow depression. In this system, there is considerable double drainage where the laterals and mains join and the system may not be very economical. However, it is particularly suitable where the laterals are long and the area requires thorough drainage.

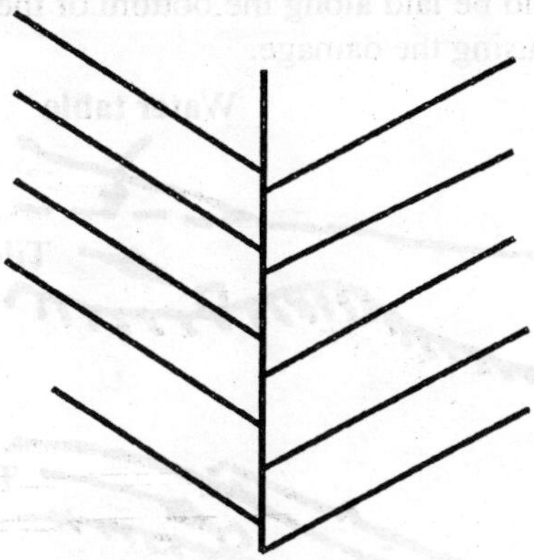

Figure 7.3. Herringbone Type Drainage System

(c) Gridiron type drainage system

The gridiron drainage system is similar to the herringbone drainage system except that the laterals enter the main only from one side. The system could be more economical than the herringbone system as less numbers of junctions are used. It is used on flat regularly shaped fields and on uniform soil. It is more economical than the herringbone system because the number of junctions and the double drained areas are reduced. Where there is a broad flat depression, which is frequently a natural water course, a main may be placed on both sides of the waterway. This system, known as double main system, is essentially to separate gridiron patterns. Placing a main on each side of the depression serves a dual purpose; it intercepts the seepage water and provides an outlet for the tile laterals.

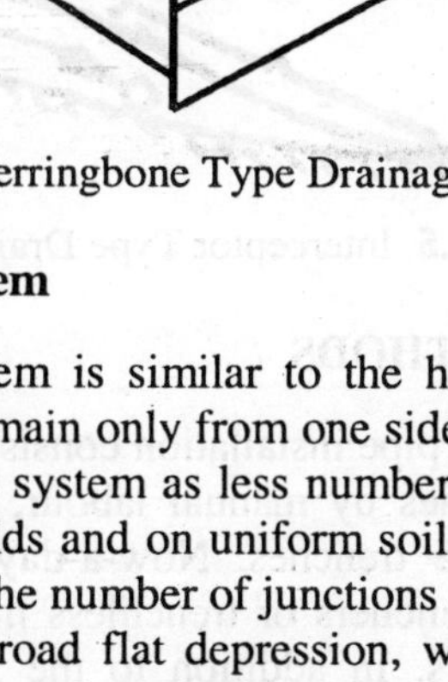

Figure 7.4. Gridiron Drainage System

(d) Interceptor type drainage system

The interceptor is placed at the upper edge of a wet area to intercept the sub-surface seepage. Open drains are also used for the purpose but require considerable land area. The interceptor or cutoff system is used to intercept seepage water from hillsides; wet areas are formed by seepage water moving horizontally through permeable layers, which lie over an impermeable layer. This condition is indicated by seepage along a horizontal plane near the foot of the slope or at a break in grade. The interceptor drain should be laid along the bottom of the permeable layer in order to intercept the seepage causing the damage.

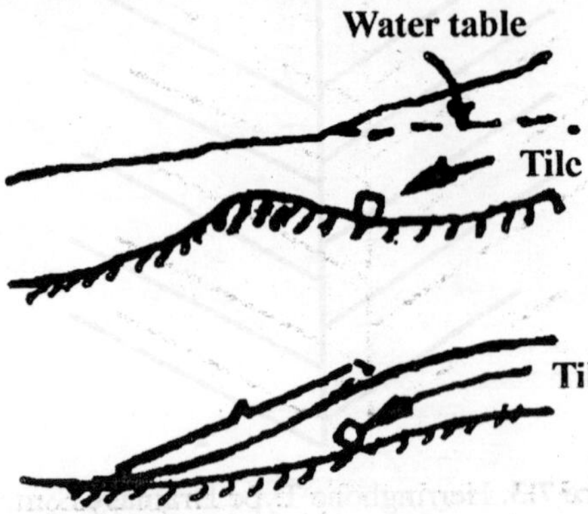

Figure 7.5. Interceptor Type Drainage System

7.2 CONSTRUCTION METHODS

The classical method of pipe installation consists of marking the alignments and levels, excavating the trenches by manual labour, placing the pipes and envelope material, and backfilling the trenches. Now-a-days, field drains are installed by drainage machines, either trenchers or trenchless machines. Concrete collectors are often installed by excavators. In addition to the mechanics of installation, other important matters are the work planning, the working conditions, and supervision and inspection.

Setting out alignments and levels

To mark alignments and levels, stakes are placed in the soil at both ends of a drain line, with the top of the stakes at a fixed height above the future trench bed. The slope of the drain line is thereby indicated. A row of boning rods is then placed in line (both vertically and horizontally) between the stakes, with an extension at the upstream end of the drain line, where the run of the drainage machine ends. The boning rods are thus in a line parallel to the trench bed. The driver of the drainage machine achieves grade control through sighting. The same principle can be applied when drains are installed manually.

The location and alignment of the drain lines must be set out before the actual digging can begin. First, the downstream location of the drain is marked off by placing a row of pegs along the collector drain at the design drain spacing. Next, the centre line of each drain is set out by placing another row of pegs at the upstream end. Stakes are placed in the soil at both ends of the drain line with the top of the stakes at a fixed height above the future trench bed using a levelling instrument. This very clearly indicates the drain line. The direction of the field drains is assessed standing at the starting point at the collector line, thereafter marking off the location of the field drains with pegs.

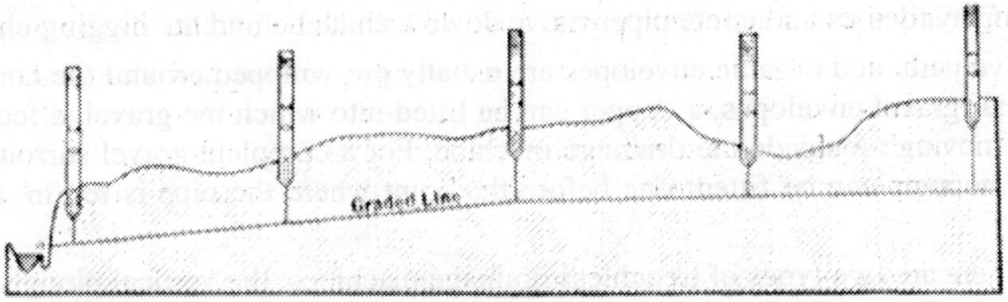

Figure 7.6. Downstream location of the drain is marked by Pegging

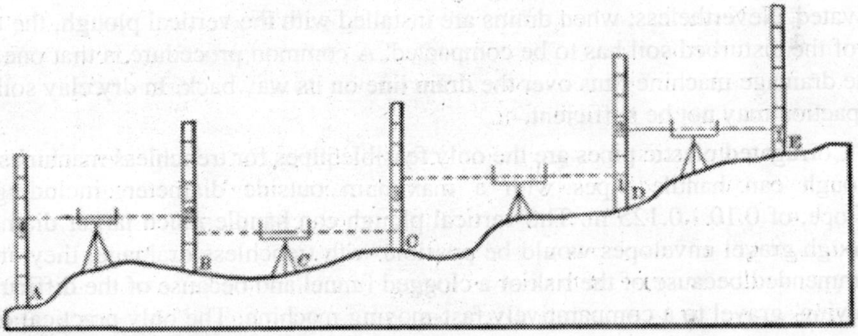

Figure 7.7. Stakes are placed for levelling

Nowadays, most drainage machines have grade control by laser. An emitter, placed on a tripod near the edge of the field, establishes an adjustable reference plane over the field by means of a rotating laser beam (Figure 5.4). A receiver, mounted on the digging part of the drainage machine, picks up the signal. The control system of the machine continuously keeps a fixed mark in the laser plane. One position of the emitter can serve the installation of a fairly large number of drains.

7.3 MACHINERY FOR DRAIN PIPE INSTALLATION

The most common types of machines for installing field drains fall into two categories: trenchers and trenchless machines. Trenchers excavate a trench in which the pipe is laid, whereas, trenchless machines merely lift the soil while the pipe is being installed.

Trenchers install the drains by excavating a trench and laying the pipe, including the envelope if applicable. The trench is backfilled afterwards by a tractor equipped with a dozer blade. Trenches should be backfilled the same day as they are dug to avoid a possible destabilization of soil under wet conditions (irrigation, rain, high water table). Running a tractor wheel over the backfilled trench, filling it up, and running over it again will take care of the required compaction. This procedure ensures that only the top part of the trench backfill is compacted, while the deeper part of the backfill retains a good permeability and a low entrance resistance.

The corrugated plastic pipe for small-diameter field drains is carried on the machine on a reel and is fed into the trench. Larger-diameter corrugated pipes (e.g. for collectors) are usually laid out and coupled in the field beforehand. The continuous tube is subsequently picked up and laid in the trench by the machine as it moves along. Clay tiles and concrete pipes move down a chute behind the digging chain.

Synthetic and organic envelopes are usually pre-wrapped around the corrugated pipe. For gravel envelopes, a hopper can be fitted into which the gravel is fed from a trailer moving alongside the drainage machine. For a complete gravel surround, two gravel hoppers can be fitted: one before the point where the pipe is fed in, and one after.

There are two types of trenchless drainage machines: the vertical plough and the V-plough. The vertical plough acts as a subsoiler: the soil is lifted and large fissures and cracks are formed. The V-plough lifts a triangular "beam" of soil while the drain pipe is being installed. Backfilling is not needed, because no trench has been excavated. Nevertheless, when drains are installed with the vertical plough, the upper part of the disturbed soil has to be compacted. A common procedure is that one track of the drainage machine runs over the drain line on its way back. In dry clay soil, this compaction may not be sufficient.

Corrugated plastic pipes are the only feasible pipes for trenchless machines. The V-plough can handle pipes with a maximum outside diameter, including the envelope, of 0.10 - 0.125 m. The vertical plough can handle much larger diameters. Although gravel envelopes would be possible with trenchless drainage, they are not recommended because of the risk of a clogged funnel and because of the difficulty of supplying gravel to a comparatively fast-moving machine. The only practical option is to use pre-wrapped envelopes.

The bottleneck for the speed of pipe installation is usually not the capacity of the drainage machine, but the organization and logistics connected with keeping the machine going. The preparation of the site (e.g. setting out, removing obstacles) is important, as is the operation and maintenance of the drainage machine (fuel supply, spare parts). In addition, the supply of pipe and envelope material needs to be properly organized.

7.4 INSTALLATION OF SUBSURFACE DRAINAGE SYSTEM

Methods of Installation

The following installation methods can be considered:

- Manual installation;
- Combined mechanical and manual installation, (installation of pipes by hand in trenches dug by excavators);
- Mechanical installation.

(a) Manual Installation

Digging trenches by hand and placing the pipes in the trenches by hand was common practice in the past and is still practiced sporadically. Special tools have even been developed for manual installation (Figure). Manual installation is a possible alternative:

(i) If the groundwater is not high (placing a pipe at a desired grade under water is practically impossible);

(ii) If the depth is limited (<1 m, greater depth is only feasible under certain conditions with stable soils and seasonal water tables below drain depth).

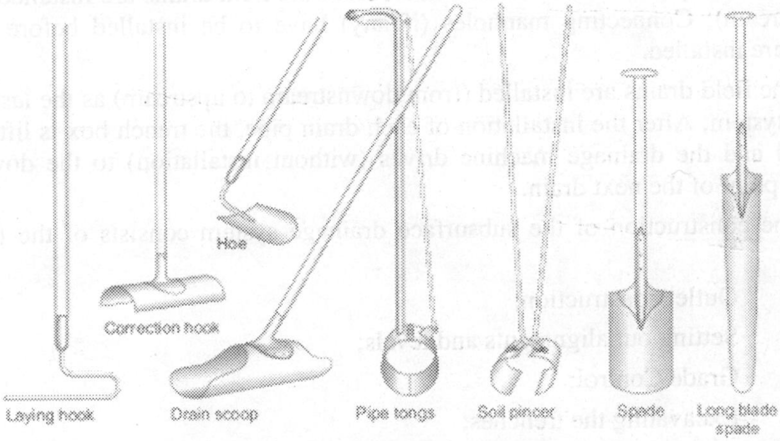

Figure 7.8. Tools to install drains by hand

(b) Combined Mechanical and Manual Installation

The combination of mechanical and manual installation consists of digging trenches with hydraulic excavators and placing the drain pipes into the trench by hand . The excavation of a uniform graded trench bottom on which the pipe can be laid with the proper grade is difficult and the progress of the work is often slow. The depth and grade control under those conditions has to be done with the aid of levelling instruments; consequently the precision is very much dependent in the skill of the operator.

A laser system can also be used, but this will not automatically regulate the digging depth of the excavator. Proper installation can only be done if there is no water in the trench and the soils are stable. Under unstable soil conditions installation with an excavator is practically impossible and/or can be dangerous for the labourers.

(c) Mechanical Installation

Mechanical installation consists of installing the drains either with a trencher or a trenchless drainage machine. With this method drains can be installed in most conditions with an automatic depth and grade control, either above or under the water level. Drainage machines are expensive pieces of equipment.

7.5. PLANNING AND PREPARATORY ASPECTS FOR DRAINAGE INSTALLATION

The installation process of pipe drainage systems should fulfill a number of conditions to assure that the installation is cost effective and results in systems that are functional.

Steps in pipe drain installation

For planning and preparation it has to be kept in mind that drainage systems have to be installed starting from the downstream side towards the upstream side.

Collector pipes have to be installed before the field drains are installed (starting downstream); Connecting manholes (if any) have to be installed before the field drains are installed.

The field drains are installed (from downstream to upstream) as the last element of the system; After the installation of each drain pipe, the trench box is lifted out of the soil and the drainage machine drives (without installation) to the downstream staring point of the next drain.

The construction of the subsurface drainage system consists of the following steps:

1. Outlet construction;
2. Setting out alignments and levels;
3. Grade Control;
4. Excavating the trenches;
5. Placing the drain pipes;
6. Placing the envelopes;
7. Installation of the junctions/manholes;
8. Backfill of the trenches.

Outlet construction

Gravity outlets

If gravity outlets are used, the side slopes of the open drains must be protected from erosion by the out flowing drainage water.

Pumped outlets (sumps)

In flat or gently sloping areas, the land gradients are generally not sufficient for free fall and gravity disposal of drainage water into surface drains. Pumped outlets are established by constructing a sump at the point of disposal of the drainage system and by installing a pumping system to pump water and maintain gravity flow within the system. The size of the sump depends on the area drained by the collector unit and the operating time of the pump (pumping capacity).

Excavating the trenches

There are three methods of mechanical installation:

- ***Excavator.*** All steps in the implementation process are separate steps, implemented one after the other. The trench is dug by the excavator to about 5 cm above the required drain depth and up to the last few centimeters, when the leveling and placing the drain pipes is done by manual labour;
- ***Trencher.*** Digging the trench, placing the drain pipes and (if applicable) the envelope, is done in a one-time operation. The pit from where the laying of the pipe will start is either dug by the trencher itself or an excavator;
- ***Trenchless.*** Just like the trencher method, it is a one-time operation, but instead of digging a trench the pipe is directly ploughed into the soil.

Placing the pipes

Several methods can be used to place the drain pipes, depending on the type of pipe:

- Concrete/clay drain pipe. The drain pipes are loaded to a platform on the machine and then put along the chute in the trench box to the bottom of the trench. This requires one labour on the platform to put the pipe in the chute and one labour in the trench box to put cloth or other sealing around the joints;
- Flexible corrugated drain pipe. The field drain pipes are delivered in coils and the coils are put on reels attached to the machine. The drain pipe is guided over rollers into the trench box. A press pulley puts the pipe at the bottom of the trench;
- Plastic collector pipes are larger in diameter and cannot be coiled. The pipes are delivered in sections of 6-12 m. These larger diameter pipes are usually laid out on the field beforehand. The pipe sections need to be connected in the field over the full length of the collector drain before the pipe laying starts and then guided through the machine.

Placing envelope

Synthetic and organic envelope material can be pre-wrapped around the pipe, this is usually done in the pipe factory. Envelopes are either applied in voluminous layers (> 10 mm) completely surrounding the drain pipe in bulk or as a pre-wrapped mat, or as thin sheets. A granular envelope requires a considerable fleet of extra equipment to ensure a continuous supply to the machine.

Installation of junctions or manholes

The field drains are connected to the collectors by means of standard drain pipe fittings (cross pieces, T-joints, Y-joints) or through junction boxes or manholes. Installation of the manhole or field-collector pipe connection is done at the start of the installation of the field drain pipes, after the collector has been installed.

Backfilling of trenches

Backfill of the drain trench is a three-step operation:

- **Blinding.** Careful placing of an initial backfill of 0.15 to 0.30 m of soil around and over the drain is referred to as blinding. This is done to ensure that the drain will remain in line when the remaining excavated material is placed in the trench. Blinding the drain may be done by shaving off the topsoil at the top of the trench with a spade or with an attachment (scraping knife) to the trench box. Care should be taken that the alignment of the drain is not changed;
- **Backfill.** The fill should be firm but not compacted too much so that it prevents the passage of water to the pipe.
- **Compaction.** Compacting is required to avoid serious problems arising in irrigated areas when water moves rapidly through the unconsolidated trench fill causing severe erosion (piping).

Trench backfilling is done by the following methods:

- Hand with shovels;
- Bulldozer;
- Grader;
- Tractor equipped with a dozer blade;
- Screw augers mounted on the trenching machine.

Site clean-up

Surplus soil that is not injurious in nature should be spread over the surrounding field. Material such as large stones and roots that are likely to damage implements or livestock, or of a size and character abnormal to material found on the surface of the field, should be removed.

Figure 7.9. Subsurface Drainage Pipe

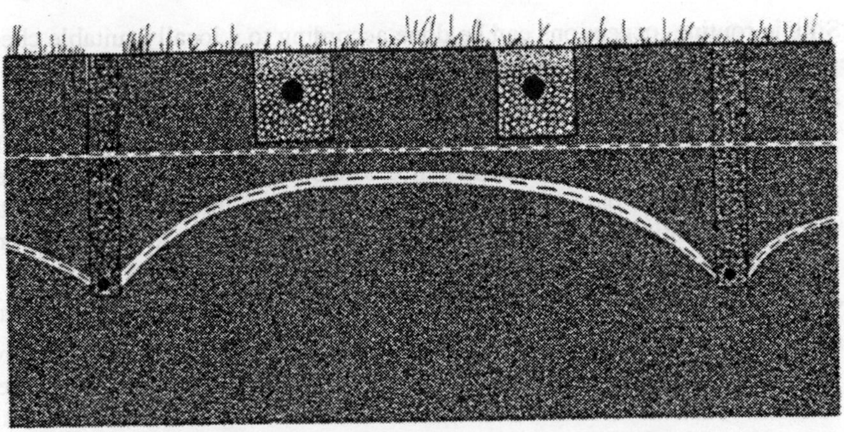

Figure 7.10. Subsurface Drainage Section

7.6 SUPERVISION AND INSPECTION DURING INSTALLATION OF DRAINAGE SYSTEM

Inspection and supervision of drainage system during installation are required:

- To ensure that design specifications are complied with;
- To handle unforeseen conditions during installation;
- To check the quality of the materials used (pipes, envelope), which includes certification and a site-check on possible damage during transport and handling;
- To ensure good workmanship, including the proper alignment of pipe lines, which should be straight and according to the design slope, within an accepted tolerance (e.g. half the inside pipe diameter for field drains), and with proper joints;
- To see that the trenches are properly backfilled and compacted;
- To assess the need for any extra work or modifications, which implies that the supervisor should be a well-qualified person?

7.7. MONITORING

There are three kinds of checks to be made of pipe drain systems: a post-construction check, routine checks, and thorough checking. A post-construction check is done to ascertain whether the construction was done to an acceptable standard, and whether the drainage works have been delivered in good functional order. This check is mainly covered under the field supervision. Routine inspections are simple operation-and-maintenance checks to verify whether the system is functioning properly, and to see whether there is any need for repair or cleaning. Some items covered by this type of inspection will be listed below. A thorough check of the functioning of the system may follow after routine inspection has revealed significant problems. Such a check may also be intended as a monitoring program, aimed at improving the empirical basis for future drainage projects in the region.

Simple routine inspections can be done according to a locally suitable checklist. Generally important points to include in such a list are:

- Check the drainage base (i.e. Check whether the pipes have free outflow, especially in a period when drainage is most needed; note, however, that an occasional, very brief submergence of the outlets is normally accepted). A good drainage base is the first and foremost condition for a pipe system to function satisfactorily. If the drainage base is found to be unsatisfactory, the main drainage system should be maintained or improved;
- Check that pipes are discharging during and shortly after rain or irrigation;
- Monitor water levels in manholes. High water levels in an individual manhole indicate an obstruction in the collector. When high water levels are found, the water levels in all manholes along a collector should be compared, which may give a clue as to where the problem is;
- Check whether sediments or other pollutants have accumulated in the silt traps of manholes;
- Look at the land surface for wet spots, as signs of waterlogging, a few days after rain or irrigation;
- Check the depth of the watertable, especially where wet conditions are found;
- Look for any damage to pipe outlets and manholes.
- Note that the observations on pipe outflow, water levels, and wet field spots should, of course, concern the same drainage event and the same drain pipe.
- A suitable time schedule for the above routine inspections would be to start with a first inspection shortly after the system has been installed, during the first or second drainage event when the drains should be running. Further inspections could follow about once a year, a frequency, which after a few years without problems, could possibly be reduced to once every two years.

Thorough checking

A thorough check of the functioning of the system may follow after a routine inspection has revealed significant problems. Such a check may also be intended as a monitoring programme, aimed at improving the design of future drainage projects in the region.

Beside these checks on the physical performance of a drainage system, the effectiveness of the investment in drainage should also be assessed; in other words: "Is the drainage system working as designed?". A monitoring and evaluation (M & E) programme could make such an assessment and could be used to check the criteria used for the design. Monitoring and evaluation should usually be considered from a long-term viewpoint, and should be based on factors that are relatively easy to evaluate. Consideration should be given to the proper collection, storage and retrieval of data. This is of the utmost importance for the subsequent physical and economic analysis of the project.

7.8. MAINTENANCE OF DRAINAGE SYSTEM

The maintenance of drainage system required during and after the installation of the system. It would be useful for you to know the causes of pipe drain failure so that care can be taken. The reasons for the same are: lack of inspection or maintenance, improper design, improper construction, and lack in manufacturing process and materials used. In the de the major problems of failure are: insufficient capacity, pipe placed too near the surface, and lack of auxiliary structure such as surface inlet. The preventive maintenance includes: Open ditch outlets should be kept free of weeds and tree growth. End of outlet pipe should be covered with a flood gate, a screen, or a flap gate. The outlet ditch should be free from sediments. Pipe drains should also be kept free from sediments. During the first year, drain lines should be carefully watched to detect evidence of failure. The corrective measures includes: Pipe drains that become filled with sediment or plant roots may be cleaned with a suitable sewer rod; Where sufficient water and grade are available, sediment can be washed out. Water can be forced at high pressure through a hose inserted from the outlet end.

It is requires that a good drainage base is maintained, that the system is inspected regularly and that repairs and cleaning are done when necessary. Operation and maintenance should have available the data on the drainage works. These should include an accurate map of all main components like field drains, collectors, connections and outlet structures. Also the elevations of collector points (outlet, inflow and outflow levels in manholes, longitudinal sections), of field drain outlets, and of reference points on major structures like manholes should be include. For future flushing operations, it also needs to know the exact location of blind junctions between field drains and collectors.

Once a drainage system has been installed, arrangements have to be made to ensure that it will function properly for a long time to come. For this, a good plan for its operation and maintenance is needed. In this plan, the following issues have to be addressed:

- Why and when is maintenance necessary? What are the objectives? What will be the frequency? What are the costs and benefits?
- Who is responsible for the planning, execution, control, and financing?
- How is it to be done and by whom: the government, the farmers, or contractors?
- How are the costs to be financed? Who will pay and how is the money to be collected?

Structures associated with pipe systems or buried pipe drains may include:

- Inlets
- Outlets
- Silt and sand traps
- Standpipes
- Pressure relief/air inlet valves
- Manholes

- Crossings
- Pumping stations

Problems associated with these structures and their maintenance and repair requirements are generally similar. General maintenance due to age, natural attrition, and design or construction inadequacies are important.

Pipe Systems

In pipe distribution systems the maintenance of the conduits as well as the many appurtenances, such as gates, valves, metering devices, etc., must be considered,. The maintenance of some of these items, such as pumps, motors, electrical controls and automation, are discussed later. Problems include:

- Damage to linings and coatings;
- Corrosion;
- Separation of pipe joints;
- Build up of material in the pipe and appurtenances. Work plans will include programs for:
- Monitoring
- Cleaning
- Joint repair
- Lining refurbishment.

Open Drains

Drains generally suffer rapid deterioration in condition affecting performance levels, and require comprehensive maintenance programs. Problems may include:

- Erosion
- Settlement
- Sloughing
- Siltation
- Vegetation
- Seepage.

Buried (Pipe) Drains

The major problems requiring maintenance include:

- Physical blockages
- Organic or biological blockages
- Chemical or mineral sealing and outlet restrictions.

Work plans will include:

- Monitoring

- Cleaning
- Root removal
- Cleaning and repair of outlet grills.

Pumping Stations and Electric Power Facilities

Pumps, motors, pumping stations, and electric power facilities are used extensively in irrigation and drainage projects in areas, such as:

- Motorized operation and automation of structure flow control devices;
- Sprinkler irrigation systems;
- Computerized management facilities;
- Pumping stations for pipeline distribution systems;
- Pumped drainage;
- Pump wells;
- Transformers, switchgear (often maintained by the power utility).

The equipment is generally specialized and comprehensive specific instructions on care and maintenance procedures are required, together with specialized training for maintenance personnel.

Maintenance of Surface Drains

Major problems in maintaining open drains may be due to erosion, settlement, silting, vegetation and seepage. Before the drainage season, drains should be cleaned; all vegetation should be removed, and side slopes and banks should be repaired when necessary. Siltation should be monitored and removed when required. The frequency depends on the local situation and no hard and fast rules can be given. The preventive maintenance that is undertaken prior to the occurrence of any failure or breakdown and corrective maintenance, which is undertaken only after a partial or complete failure takes place.

- The most important of the preventive maintenance activities is the control and elimination of excessive tree, bush, grass and weed growth.
- Before the drainage season, drains should be cleaned (e.g. with a shovel or a V-drag)
- The various methods of controlling growth of vegetation are grubbing, underwater cutting, chemical-spraying, burning when dry and flame-burning when wet.
- Grazing by animals may serve a useful purpose, if dangers of trampling of side slopes out of shape, damaging of bed flooring or lining are not serious.
- Bank erosion and sediment deposition may be caused mainly at the spoil banks; these can be made flatter and also stabilized with short grass germinated through proper seeding.
- The grass so grown on the spoil banks may prevent their getting washed away easily.

- Weed and silt clearance either by manually or with machines, restoration of shape of the drain bund of a drains or its stability constitute the most important operations under corrective maintenance.
- Machines and tools used for weed clearance include mowers, scythes, sickles, brush hooks, and axes.
- When the flow velocity is high, a band of long thin reed like hand-saws may be sunk to the bottom and moved back and forth to effect cutting vegetation stems on the bed.
- Rain waters that enter a drain bring with it a lot of sediment and debris from the neighbouring areas, contributing to this run off. If a drain serves only to dispose off excess surface waters, silting of the bed can be tolerated up to even say 0.3 m.
- Also the minimum bed width to employ machines for silt clearance should be about 2 m. Drag line excavator-bucket capacities in the range of 0.4 m^3 to 1.5 m^3 are found convenient.
- The caving in of the banks is due to excessive values of bed slope as well as side slopes.
- Increasing the resistance against erosion by adding masonry rip rap, willow mattresses or brush retards may also be considered wherever appropriate.
- Siltation should be monitored and removed when required.

The major causes for the failure of open channels and drains

a. Poor and improper design: A poor design may arise due to poor location and alignment of a drain, inadequate width and depth, inadequate culvert and bridge waterway discharge capacity, or underestimation of sediment load carried by the water.

b. Improper construction: Poor construction of bed slope, bed and side surface areas might lead to: sedimentation in the channel reaches, excessive growth of vegetation, and channel and bank erosion.

c. Absence of adequate maintenance.

Maintenance of Subsurface Drains

- The design of subsurface drainage is based on soil properties, rainfall and the resulting design discharge that can be expected, the site topography and land use and the elevation of possible outlet locations, impervious structure, etc.
- The depth of a drain, from ground surface, varies between 0.75 m and 2.5 m. The lower the permeability the smaller would be the depth.
- In heavy clay soils, the drain spacing have been between 5 and 15 m. In very permeable soils, spacing can be upto 150 m.
- The gradients adopted for a drain could be as steep as 1 in 30 or as flat as 1 in 1000.

- The danger of clogging by sedimentation in flatter gradient should be guarded against.
- For the maintenance of pipe drains, the problems may be physical blockages, organic and biological blockages, chemical or mineral sealing, and outlet restrictions.
- Before the drainage season, a visual inspection of all outlets should be done, and water levels in manholes should be monitored to check for obstructions or siltation in the pipe sections.
- Pipe lines can be cleaned with specialized flushing machines which remove sediment from the pipes

Maintenance of Land surface

- The benefits derived from land forming will often depend on good maintenance in the subsequent years.
- A bedding system requires regular maintenance.
- Care should be taken to eliminate any obstructions to flow or low points in the drains because they will cause standing water and loss of crops. In graded fields, the land should be smoothed each time a field has been ploughed.
- A small leveller or plane powered by animal traction or a farm tractor can be used for this purpose.

Maintenance of Structures

- Structures normally have a higher safety factor than the drains, and in general need less maintenance.
- Regular and timely inspection is required to identify problems and maintenance needs.
- The moving parts in doors and gates should be checked for wear and tear, and inflow and outflow openings should be cleaned of debris and checked for scouring and damage to banks and the structure itself.
- In tidal areas and in rivers and drains with high sediment loads, regular flushing and/or dredging might be required.
- Without maintenance, a drainage system will not function properly and no sustainable agriculture can be achieved.

CHAPTER - 8

SPECIAL DRAINAGE SYSTEMS

8.1. MOLE DRAINAGE SYSTEM

The mole drain also known as pipeless drain in used heavy soils of low hydraulic conductivity often require very closely spaced drainage system for satisfactory water level control. The moling is a temporary method of drainage. The mole drainage is low cost due to no use of pipes and they can be installed economically at very close spacings. The disadvantage of mole drainage is its restricted life, but, providing benefit-cost ratios are favourable, short life can be accepted. The success of a mole channel and is dependent upon satisfactory water entry into the mole channel and upon the mole channel it remaining stable and open for an acceptable period. Mole channels usually range in diameter from 50 to 100 mm. The small sizes are more stable than the larger ones. The depth of mole channels varies from 0.5 to 1.2 m depending on the depth of a stable soil layer. Spacings will range from 1 to 50 m. Length of drain is usually less than 500 m depending on the grade, which may range from nearly level to 5 per cent.

Mole drains are constructed using a mole plough, which comprises a cylindrical wooden block or foot attached to a narrow leg, followed by a slightly larger diameter cylindrical expander. The foot and expander form the drainage channel and the leg generates a slot with associated soil fissures, which extend from the surface down into the channel. Mole channel walls become smoother as soil-water contents increase, the following expander increasing the smoothing effect. At high water contents in low density soils, the expander tends to seal off the connection between the leg slot and the mole channel. The water content at the time of moling should be as high as possible, provided the soil can support the tractor. A general guideline in this respect is that the soil moisture content should be between Liquid Limit and Plastic Limit. Besides the moisture content, the bulk density of the soil is also an important factor governing the power requirement and the crack formation while drawing the mole plough through the soil. The success of a mole drainage system is dependent upon satisfying two requirements: achieving the desired water flow path for the particular drainage situation, and installing stable mole channels. The installation technique adopted must meet these requirements. If the soil above the mole drain is less pervious, sub-soiling helps in developing wider path for the shallow sub-soil water to flow towards the mole drain.

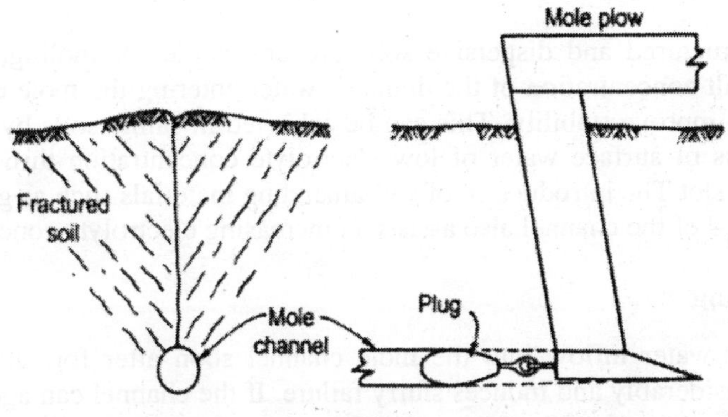

Figure 8.1. Mole Plough and Mole Channel

8.2. STABILITY OF MOLE CHANNEL

The life of mole channels depends upon soil type, soil conditions at moling, installation technique and equipment, and the subsequent weather patterns. In areas where moling is regularly practised, the working life of the channel usually varies from three to five years or more. The required life in other situations will, however, depend upon benefit-cost ratios and the ability to re-mole an area at an appropriate time after collapse. Channel life is very dependent on the magnitude of the cohesive bonds that develop in the channel-wall area following installation, and on the disruptive stresses, usually due to soil swelling, which may ensue. The stronger the bonds and the weaker the disruptive stresses, the longer will be the channel life. Mole channels collapse in numerous ways depending upon the circumstances prevailing. The roof collapse due to inadequate cohesion between soil structural units, or resulting from soil disturbance caused by the expander at installation; The wetting and drying cycles causing repeated swelling and shrinkage; it is common in swelling clays in areas where significant drying regularly occurs.

8.3. FACTORS AFFECTING MOLE CHANNEL STABILITY

Clay Content

Soil clay content is particularly important because of its influence on the magnitude of the cohesive forces, which are necessary for channel stability. Some of the best moling soils have clay contents in excess of 45%, whereas soils with less than 30% clay are rarely good molers. Uniform soils are most satisfactory. Mole channels in soils containing sand or silt pockets are more prone to collapse.

Wetting and drying

Soil structural stability on wetting is a crucial factor influencing channel life. The more resistant the soil structural units are to collapse on wetting. Structural stability is assessed on a basis of differences between the water-release characteristics of initially air-dry samples, wetted very rapidly by flooding, and slowly through capillarity.

Soil salinity

Weakly structured and dispersive soils are unsuitable for moling, unless the electrolyte or salt concentration of the drainage water entering the mole channel can be increased to improve stability. This can be achieved in saline soils by preventing the direct access of surface water of low electrolyte concentration through the leg fissures and leg slot The introduction of soil-amending materials such as gypsum into the neighborhood of the channel also assists an increasing electrolyte concentration.

Timing of Moling

Significant water inflow into the mole channel soon after formation reduces channel life considerably and induces slurry failure. If the channel can age or mature for a number of weeks (2 to 3 minimum) before being wetted, it will be much more resistant to collapse. Moling during a drying period is therefore, most desirable. An extended drying period after moling in smectitic clay soils may, however, cause excessive and undesirable shrinkage in the leg slot area, leading to considerable channel in-fill with surface soil. Where moling has to be done in the presence of a perched watertable or of free water at moling depth with immediate water entry, this moling should be regarded as a de-watering operation and the operation should be repeated later under drier conditions.

Moling Depth

Mole channels are commonly installed at depths between 0.4 and 0.7 m, although there is no restriction to deeper installation. Wherever, possible, installation should always be in the most suitable and structurally stable soil layer. Moles should be installed deeper in situations where deep soil drying and cracking are likely to occur; this reduces the risks of cyclical swell/shrink failures.

Mole Drain Spacing

Because of the semi-permanent nature of mole drains and the risks of collapse, spacings closer than those required to meet the drainage design criteria are usually adopted. This ensures that drainage performance is not seriously impaired if some of the mole drains collapse. These closer spacings have minimal implications on cost, because mole drains are very cheap to install. Common mole drain spacings range between 2.0 and 3.5m.

Length of Mole drain

The length of the mole run between outfalls is dependent upon the estimated chances of moling success. The greater the risk of channel failure, the shorter the length of run adopted, to minimize the effect of local collapse on overall drainage efficiency. Factors such as the moling qualities of the soil, surface irregularities and gradients possible reverse gradients, and the presence of sand and silt pockets are usually considered when deciding upon run length. Common lengths of run vary from approximately 20 to 100 m. The length may have to be reduced on steep gradients to avoid channel erosion.

Diameter of Mole Foot and Expander

The choice is dependent upon the most probable type of mole channel failure, soil conditions at installation, depth requirements, and the power available. Where unconfined swelling failures are expected, larger diameter mole channel should be installed to extend their working life. Soils prone to roof collapse and expander failure, on the other hand, benefit from smaller diameter channels. Expander failure is very common following installation under marginally dry conditions. In these circumstances, the expander diameter should be only slightly larger than the foot diameter, its function being only to smooth out the channel walls rather than induce further deformation.

In situations where it is necessary to seal off the mole channel from the leg slot above, this can be achieved by increasing the expander diameter relative to the foot diameter. Conversely, in circumstances where a good open connection is required between leg slot and channel and the leg slot and fissures are only weakly formed removing the expander completely will help keep the connection open. This latter situation is common when moling under softer, very plastic soil conditions.

Outlets

A good stable outfall for the mole channel is essential for successful moling. Outfalls into open ditches must be stabilized with short lengths of pipe (0.8 to 1.0 m long) at the outlet.

Slope of Mole drain

The excessive channel grade may cause an erosion failure and, on sloping areas, grades between 1 in 40 and 1 in 60 are often considered optimum. Lower grades are satisfactory, providing local reverse grades can be avoided. Extended water ponding within the channel, particularly to depths greater than half the channel diameter, rapidly increases the rate of channel collapse. Reverse gradients, a major cause of ponding, therefore have to be avoided.

8.4. VERTICAL DRAINAGE

Drainage by wells is a form of sub-surface drainage. It is also referred to as vertical drainage. The vertical drainage consists of pumping, from a series of wells, an amount of groundwater equal to the drainable surplus. The success of tube well drainage depends on many factors, including the hydrogeological conditions of the area, the physical properties of the aquifer to be pumped, and those of the overlying fine-textured layers. Enough water has to be removed from the aquifer to produce the required drop in hydraulic head, and, for vertical downward flow, the hydraulic conductivity of the overlying layers must be such that the water table in these layers responds sufficiently quickly to the reduced head in the pumped aquifer. Another important factor, of course, is that skilled personnel are needed to operate and maintain the tube wells, and to monitor water tables and the quality of the pumped water.

Subsurface drainage of agricultural land can be done by horizontal pipe drainage systems, but when the aquifer is deep drainage by wells (vertical drainage) may be a feasible alternative because the well spacing can be quite wide achieving the same effect on the lowering of the water table.

The law of Darcy states:

$$\frac{\delta J}{\delta X} = -\frac{Vx}{Kx}$$

where:

J: the level of the water table at distance X, taken with respect to the level of the impermeable base of the aquifer (m)

X: a distance in horizontal direction (m)

Vx: the apparent flow velocity at X in horizontal X-direction (m/day)

Kx: the horizontal hydraulic conductivity (m/day)

δX: a small increment of distance X (m)

δJ: the increment of level J over increment δX (m)

δJ/δX: the gradient of the water table at X (m/m)

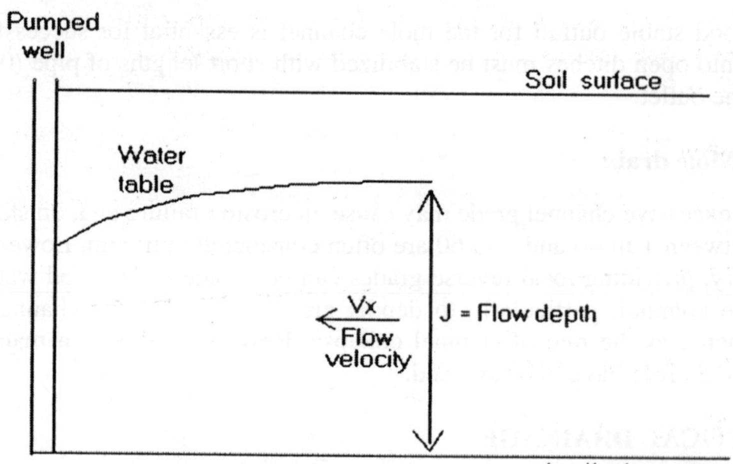

Figure 8.2. Flow of Water toward the Well

The velocity of flow to a fully penetrating well, i.e. a well reaching the bottom of the aquifer, in a homogeneous aquifer can be found from the water balance as:

$$Vx = \frac{\pi\left(Z^2 - X^2\right)R}{2\,\pi\,X\,J}$$

where

R: the long term average (quasi steady) recharge by downward percolating water stemming from rain or irrigation water equalling the long term pumping rate from the well (m/day)

Z: the radius of a circle around the well to where the influence of the well extends, the zone of influence m)

The factor $\pi(Z^2-X^2)R$ represents the amount of water (m^3/day) stemming from the surface of a segment between the circles with radius Z and X, and the factor $2\pi X.J$ is the cylindrical cross- section (m^2/day) of flow at distance X.

When the wells are arranged in a square pattern we may equate the surface area under the radius of influence of the well with the surface area geometrically associated with a well:

$$\pi Z^2 = L^2$$

so that

$$Z = 0.56\ L \text{ and } L = 1.77\ Z$$

For a regular triangular pattern we find

L = 1.73 Z and L = 0.58 Z. The difference is relatively small.

When the well pattern is somewhat irregular we may use: $L=0.57\sqrt{(A/N)}$, where A is the area served by N wells, and $Z=\{\sqrt{(A/N)}\}/1.75$ using factors intermediate between those found before.

8.5 SUITABILITY OF VERTICAL DRAINAGE

Drainage by wells is feasible only under certain conditions mentioned below:

Aquifer Conditions

The area to be drained should be underlain by an aquifer, pumping from which will adequately lower down the water table. The water table should be continuous from the shallow layer down to the deep pumped layer.

Water Quality

The underground water should be of satisfactory quality. Thus the pumped water could be used for irrigation.

Soil Condition

The hydraulic conductivity of the soil layers to be drained is important as these soil layers should allow free percolation of water. There should not be any intervening layers of low hydraulic conductivity impeding the movement of water. Drainage wells are most successful in irrigated areas when the pumped water could be used for irrigation. A water balance could be obtained keeping the water table at the desired level.

8.6. BIO-DRAINAGE

The trees or crops having fast transpiration rate use for drainage of land. Bio-drainage is removal of excess water with effects of the high rate of withdrawal of groundwater by certain plants (e.g eucalyptus, poplar). The transpiration capacity of a plant depends upon its species, root depth and spread, canopy area, leaf area and leaf structure. When the transpiration is met primarily by drawing groundwater, the process is called as bio-drainage. This technique is applicable to physically and chemically degraded lands. It is then an alternate use, rather than reclamation of such lands for normal crop production. The major advantage of bio-drainage is a low-cost measure, does not require gravity outlet, operation and maintenance are required only at the plant establishment stage, no energy for operation and supply of fuel and fodder material. The disadvantages of bio-drainage are some area is needed for growing the plantation, which cannot be available for crop cultivation and good quality water should be available for plant establishment. The perspective sites for tree plantation for bio-drainage are government lands and fallow lands with low productivity.

In case of bio-drainage, the area within the cluster of the plants cannot be put to use for normal crop production. The major difficulties will be lack of light for photosynthesis, in making appropriate fields without disturbing the plants, in laying out irrigation distribution system and operation of farm equipment.

CHAPTER – 9

SUPPLEMENTARY PROTECTION STRUCTURES

There can be some reasons to reduce drainage temporarily (e.g. environmental considerations, unwarranted and harmful leaching of fertilizers in winter, supplementary irrigation and special water regimes for rice and other crops). Devices for controlled drainage can be installed in open ditches or on subsurface drains. Unperforated pipes with a length of 5 m, leading drains into or from the control structure, should be used to prevent seepage around the structure. Very simple control tools can be used such as an elbow or plug with a riser or a plug with a bypass. Structures with crest boards are common in open ditches. Very sophisticated structures with crest boards, floats or electric water level sensors in a sump, either located on the drain line or midway between drains, can be used as well (Madramootoo *et al.*, FAO, 1997; Schwab and Fouss, 1999). Simple yet reliable control devices can be made locally, however, with available means. Control structures are made of masonry, cast-in-place concrete or pre-cast segments.

Drainpipes serving both drainage and irrigation purposes are sometimes laid without slope. However, this is not necessary as long as the gradient remains sufficiently small. Automatic controls are required to maintain the water level at the drainage outlets, which serve as inlets for subirrigation systems. Subirrigation should not be practised in arid regions where soil salinity is a potential problem.

9.1 INLETS

Inlets are to be built into the drain in places where surface water is likely to accumulate. A surface inlet is an intake structure for the removal of surface water from potholes, road ditches, other depressions and farmsteads. In some tight soils, water tends to collect in low spots and it may be desirable to provide surface inlet to let the water get to the tile faster than it will pass through the soil. In view of the sedimentation risk involved, surface water inlets are not very common. Surface water should preferably be evacuated through a network of open drains. Two possible types are blind inlets and open inlets.

Blind Inlets

Blind inlets are intended to drain stagnant pools, while sediments are intercepted. They consist of a trench above a drain that is filled with porous material. Durable material, such as stones, gravel and coarse sand is preferred as trench backfill. The gradation may vary from finer material at the surface to coarser with depth, although the trench can also be filled with one suitable porous material. The advantage of blind inlets is the initial low costs and the lack of interference with tillage operations. However, in general the use of blind inlets has been unsatisfactory because they tend to clog at the surface with fine soil particles and other sediments.

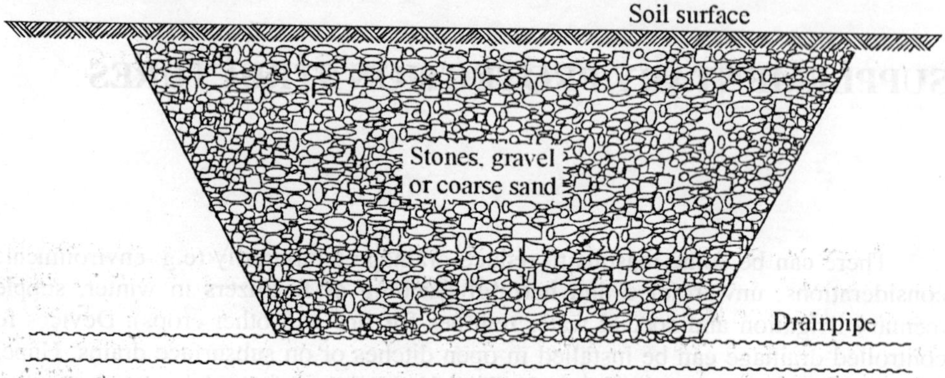

Figure 9.1. Blind inlet

Surface Inlets

Surface water inlets are incidentally used to evacuate surface water from localized areas through the drainage system when the construction of ditches is not feasible or impractical. A proper silt trap is essential to prevent or reduce drain siltation. The open inlet can be in the collector line although it is better located next to the collector and connected to it with a siphon as a safeguard against poor maintenance. Surface inlets are usually made of masonry or cast-in-place concrete, but concrete and rigid plastic pipes can also be used. A metal grating is usually installed to restrict the entry of trash and waste.

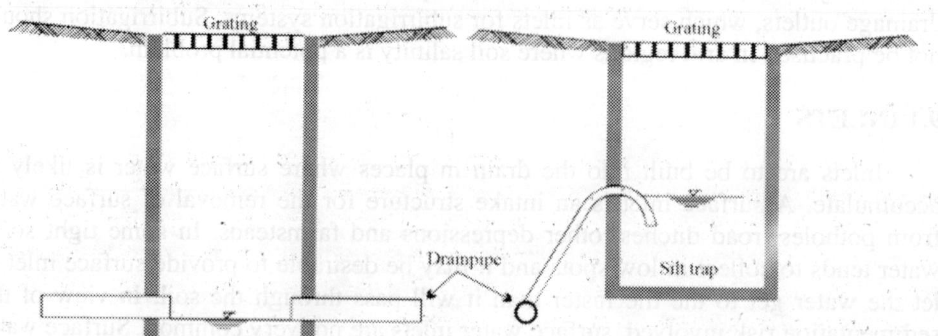

Figure 9.2. Surface inlet

9.2 OUTLETS

The lack of good natural outlets is perhaps the greatest difficulty that the farmers have to surmount in draining flat prairie. To make under-draining successful, deepen the watercourse by making open ditches, even larger and deeper than the existing. In order that a tile drain may discharge all the water that it is capable of carrying, the water must flow away with perfect freedom. It is thought by many that some contrivance may be used by which the force of gravity may be circumvented, and the drain made to discharge its water, notwithstanding a faulty outlet. A box or barrel is sunk at the lower end of the drain, and the water made to rise twenty or thirty inches and then flow away through a shallow open ditch. The effect of this is shown in figure. We must remember that if water will enter at the joints of the drain, it will flow out into the soil just as readily if the stream in the tile is retarded or entirely stopped. The water, as shown in the figure, flows through the drain to its outlet A, whereby being held by the sides of the pit or box, the water must rise to the level Bb before it can flow off. The soil above the drain becomes saturated up to the line Bb, thus leaving a portion of land near the outlet undrained, or only partially drained, according to the relation between the height to which the water must rise and the grade upon which the drain is laid. Nor is this a small waste of land, especially where the grade of the drain is slight. We will suppose that the mouth of the drain is three feet below the surface, that the water must rise twenty-four# inches before it can flow off, and that the grade of the drain is four inches in one hundred feet. The drain will now back water for six hundred feet, and will greatly injure the land for a distance of three hundred feet above the outlet. When the grade of the drain is great near the outlet, the damage is not so serious. When the soil above the drain becomes saturated from any cause, as in the case of an obstructed outlet, the combined weight of soil and water above the drain causes the tile to act like a continuous iron pipe, the head of water above forcing the water through the pipe to a proportionate height at the outlet. This, however, is not drainage. The drain must have a free outlet, and the tiles should never be quite full in order to give perfect drainage to all of the soil through which the drain runs.

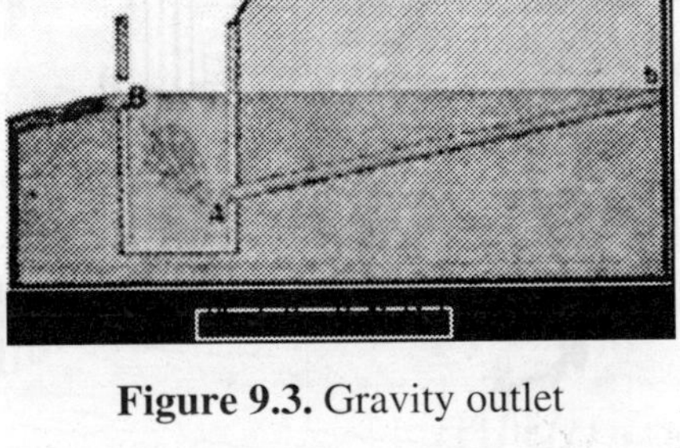

Figure 9.3. Gravity outlet

Gravity Outlets

The outlet of laterals and collectors must be protected in case of gravity discharge of the water into an open drain system. The outlet should be reliable since malfunctioning affects the performance of the entire drain or drainage system. The outlet of laterals and smaller collector drains can be protected with a non-perforated rigid pipe made of plastic, coated galvanized steel, reinforced concrete or other materials. The length of this pipe ranges from 1.5 to 5.0 m, depending on the diameter of the drain pipe, the risk of root penetration from bank vegetation and the danger of erosion under the pipe or at the discharge point. No envelope material (particularly gravel) shall be applied near the outlet and the last few metres of the trench backfill should be well compacted over the entire depth of the trench. The outlet pipe can be connected to, or slid over the drain pipe and at least half of its length should be buried.

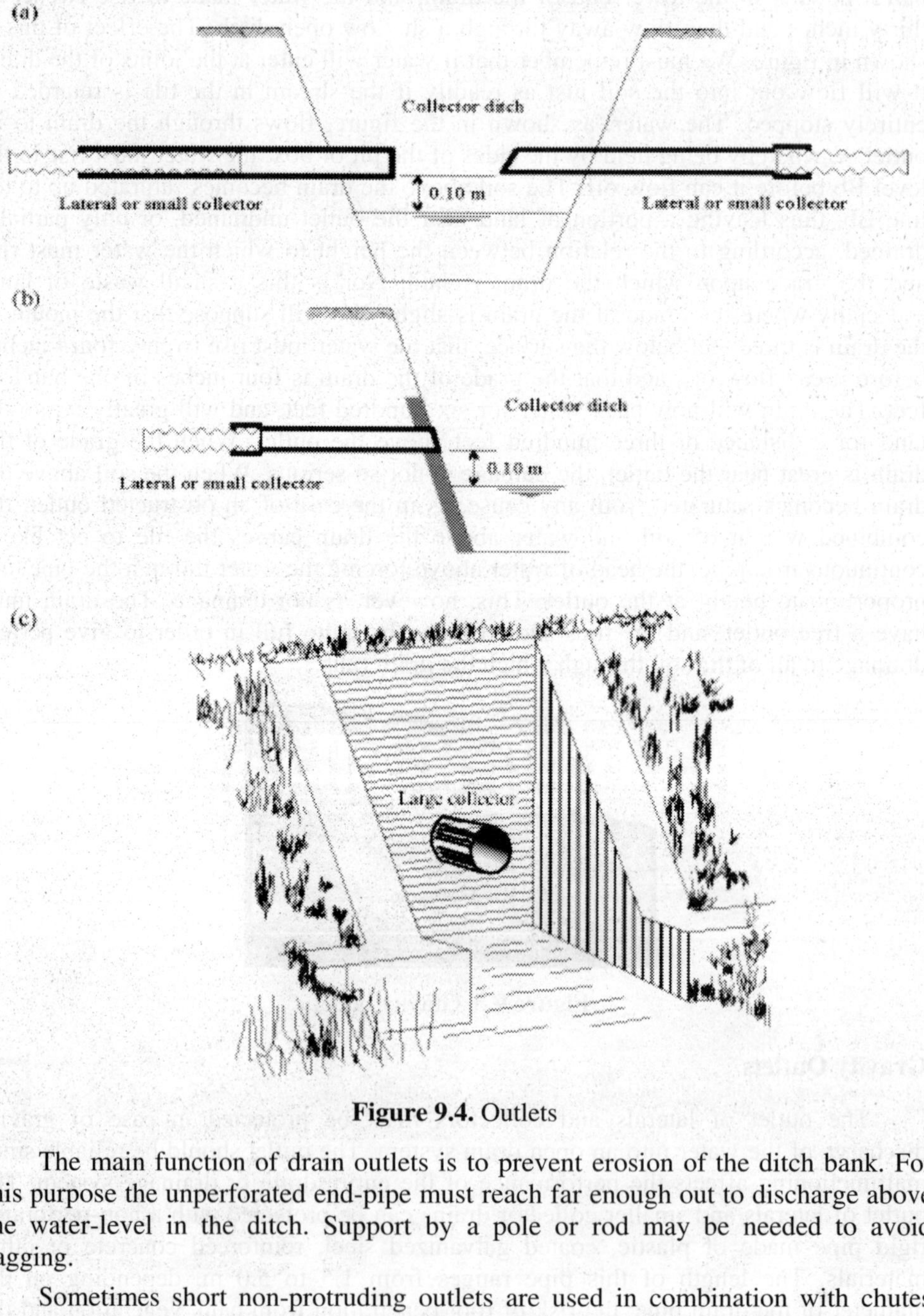

Figure 9.4. Outlets

The main function of drain outlets is to prevent erosion of the ditch bank. For this purpose the unperforated end-pipe must reach far enough out to discharge above the water-level in the ditch. Support by a pole or rod may be needed to avoid sagging.

Sometimes short non-protruding outlets are used in combination with chutes protecting the side-slope of the ditch. These chutes can be halved plastic pipes or cement gutters guiding the stream. A non-protruding pipe can also be used where there is danger of ice jams.

In spite of many efforts, no adequate solution is yet found to solve the problem of outlet interference with ditch maintenance. Plastic end pipes resist corrosion from chemicals in soil and water but burning off side slopes of ditches as a maintenance measure will be fatal.

Larger collector drains justify the use of a small concrete structure, made of masonry, cast-in-place concrete, or pre-cast segments. Outlets should be provided with a removable screen to prevent the entry of small animals. Although the outlet into open ditches may be submerged for short periods during storms, they are usually not and should be at least 0.10 to 0.15 m above the water level in the ditch at normal flow.

Pumped Outlets

Pumps are used for the discharge of water from a drainage system into an outlet ditch, when gravity outflow is not possible because of insufficient outlet depth. This situation is common with deep drainage systems that are designed for salinity control in arid and semi-arid regions. In other areas they may be needed because of insufficient outlet levels. Collector lines discharge into a storage sump with concrete base, where a float-controlled pump periodically empties the sump. Pumped outlets are more expensive than gravity outlets, not only because of the initial cost of equipment, but also due to costs associated with maintenance and power consumption.

Pumped outlets are equipped with a power unit (either electric motor or diesel engine), and pumps and pipes for lifting collected drainage water to a shallow gravity outlet. Small sumps can be constructed with large diameter plastic, asphalt-coated corrugated steel or concrete pipes while larger sumps shall be made of reinforced concrete rings, masonry or reinforced concrete.

9.3. OUTLET PROTECTION STRUCTURES

If the pipe line discharges into a stream the pipe should be long enough to provide a firm anchorage in the earth and extend out over the water of the stream. A firmly placed flat stone or rock or stone layers on which water will fall to prevent undercutting.

Metal Gate

Iron rods or flat metal gates placed across the outlet end of tile drains will keep out rodents and small animals from the drains. Small animals sometimes enter the tile drain, build their nests and may get themselves stuck fast in the drains, eventually blocking them.

Flap gates

It hinged at the top are used to protect the tile outlet from backwater when the level of water in the outlet ditch rises above the pipe line. Flood water carries considerable amount of sediment and if allowed to back up into the tile and becomes stationary, will drop the sediment, thus filling up the tile. The flap gate when fitted is automatic in action; when the water in the ditch is of sufficient height to prevent the water in the tile flowing out, it closes. When not, it opens and allows the water in the tile to drain out.

Headwall

A drain discharging into an open ditch should always be protected against erosion and a headwall of well-laid stones is sufficient in some cases. If surface water enters the outlet ditch at the same location as the tile, a concrete or masonry headwall is necessary. This must have a deep foundation and wing walls so that it will not be pushed over. A notched concrete headwall is needed where a grassed waterway and a tile drain discharge at the same place.

9.4. SILT BASINS

The silt basin is often a valuable auxiliary to a system of drains, but it is not used as much as it would be if its advantages were better understood. It may be described as a small well, placed either in the line of a single drain, or at the junction of several drains, and serves several different purposes. Figure gives an idea of the construction and use of the silt basin. It may be built of brick, stone, or plank, and may vary in diameter from twelve to twenty-four inches, according to the use for which it is intended. There should be a depth of twelve inches below the tile for the deposit of muddy matter. In the figure given it is shown two feet in diameter, constructed of brick, with a stone foundation. In draining it is often desirable that several sub-mains or branches should join at one place, and there unite in one line as an outlet to the whole system. This is the use of the basin, as shown in the figure here given. It permits us to unite several drains entering at different angles, without the objectionable feature of short turns, which we have before noticed. To facilitate the action of the drains, the outlet of the basin should be a few inches lower than the outlets of the lines of tile entering it.

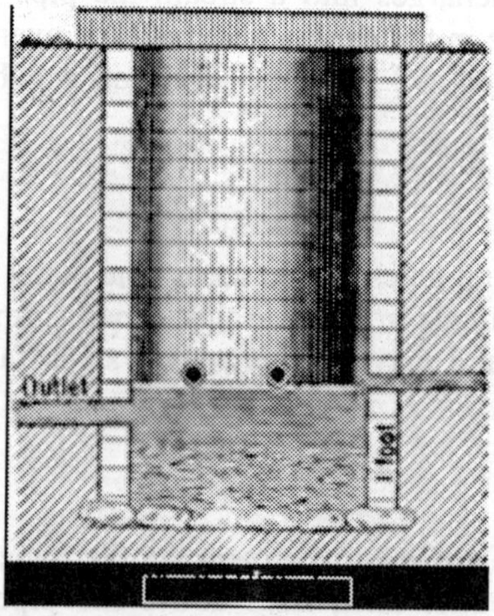

Figure 9.5. Silt basin

The fine earth, or silt, which finds its way into the tile and is carried along with the drainage water, is permitted to settle in the basin, instead of being carried on by the current, to lodge in some portion of the drain where a turn is made, or where the velocity is decreased by a less grade. The basin should have a cover, which may be removed and the silt taken out before it impedes the flow of water through the tiles.

The silt basin prevent the silt from obstructing the drain in cases where the grade suddenly changes from a steep grade to one considerably less. This retards the flow, which causes the silt coming from the upper part of the drain to be deposited at the point where the change to a less grade is made. Here is where the basin should be placed, in order that the silt may be intercepted and removed when the lower portion of the basin becomes full. For this purpose the diameter of the basin may be much less than for the purpose of collecting the water of several drains.

In the ordinary drainage upon western farms, there is but little necessity for the construction of basins for the purpose of simply collecting silt, for there is usually not enough difference in the grade to cause any alarm on that account. Yet near streams which break the land up into alternate steep slopes and flat bottoms, they are sometimes a necessity. In long mains, however, it is best to locate silt basins at various places along the line for the purpose of watching the action of the drain, and to see that its several sections are in perfect condition.

Where the soil consists of loam on a firm clay subsoil, there is very little and sometimes no deposit of silt after the drain has been in operation a few weeks. There are many sub-soils, even in prairie lands, which contain streaks of sandy material, which, for some time after the construction of the drain, will find its way into the tiles. It will be seen that the provisions made for the interception of silt must be regulated by the kind of material in the soil through which the drain runs.

9.5 BREATHERS

Breathers are installed on long lines to prevent the development of a vacuum. For slopes over 1 per cent, negative pressures could develop in a tile running full. Quantitative values could not be predicted in the field nor were the effects of these negative pressures on flow determined. Many engineers do not feel that breathers are justified.

9.6 RELIEF PIPES

Relief pipes are small-size vertical risers extending from the drain line to the surface. The riser should be made of steel pipe or sealed bell and spigot tile and should be located at fence lines where they are not likely to be damaged.

It relieves the excess water pressure in the drain during periods of high outflow, thus preventing blowouts. A relief pipe should be installed where a steep section of a main changes to a flat section, unless the capacity of the flat section exceeds the capacity of the steep section by 25 per cent.

9.7 JUNCTION BOXES

Junction boxes are used where two or more drains (laterals and/or collectors) come together or where the diameter or the slope of the collector changes. They can be pre-casted or made of masonry or cast-in-place concrete, but also rigid plastic or concrete pipes can be used for this purpose. Junction boxes can be combined with a silt trap and extended to the soil surface. The bottom of the silt trap should be at least 0.30 m below the bottom of the inlet of the downstream pipe. The invert of the entering laterals should be positioned at least

0.10 m above the top of the leaving collector to further sedimentation in the silt trap. Blind junction boxes will not hinder field works. The lid should therefore be situated at a minimum depth of 0.40 m below soil surface. They can be exposed if inspection and occasional cleaning is required. With the lid at the soil surface, the junction box is not so very much different from an inspection chamber, yet it hampers field works.

The position of blind boxes and covered manholes should be well documented. Nevertheless, finding them is often difficult. If they do not contain steel components, a lid with steel bars should be installed on top of the structure in order to facilitate easy location with a metal detector.

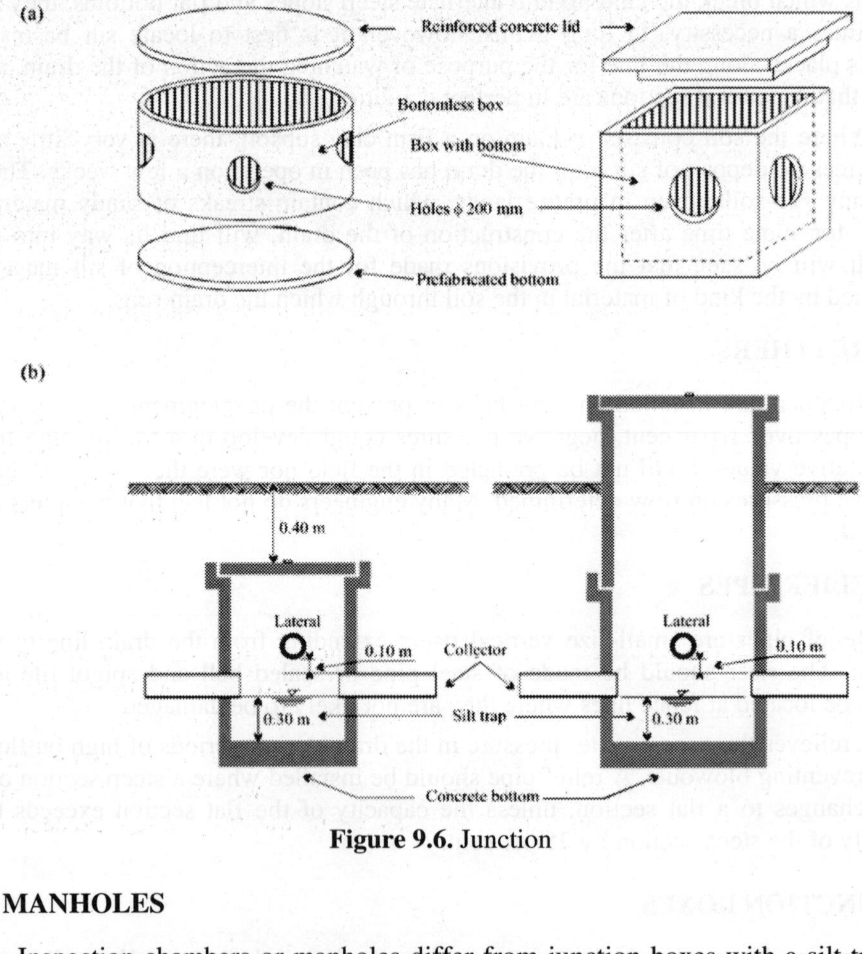

Figure 9.6. Junction

9.8 MANHOLES

Inspection chambers or manholes differ from junction boxes with a silt trap in that they provide for ready access if drains require inspection and cleaning. The material can be concrete or masonry, but also redwood has been used successfully (Luthin, 1978). Deep inspection chambers are constructed with a number of reinforced concrete rings. They should be sufficiently large and must be provided with metal rungs fixed in the wall to allow a man to descend to the drain lines. Since the lid of manholes is usually above the soil surface, they are objectionable because of their interference with farming operations. To meet this objection a capped manhole, with the top at least 0.40 m under the soil surface, can be installed with the inconvenience that the top of the manhole has to be dug out for each inspection.

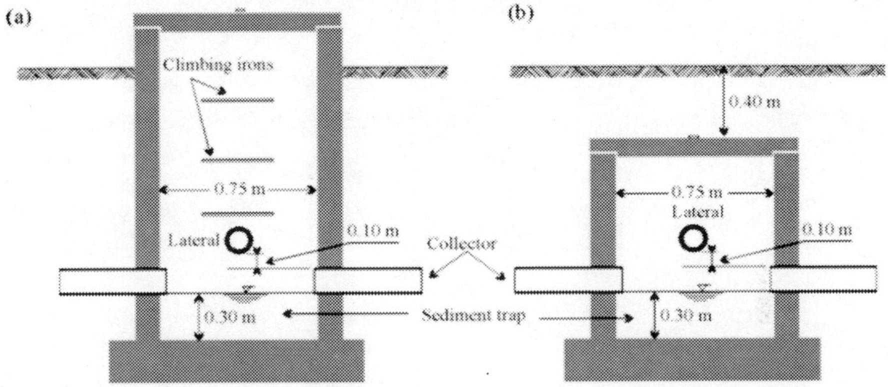

Figure 9.7. Manhole

9.9 DRAIN BRIDGES

The undisturbed natural soil in which the pipes are laid normally has enough strength to support the pipe. However, when the drain crosses a soft spot where the soil has not yet settled, e.g. a filled-in former ditch, drain bridges should be used to maintain the level of the drain during settlement of the soil. Drain bridges can be made of timber blocks on which the drain is laid or of a continuous length of solid, rigid pipe surrounding the drain.

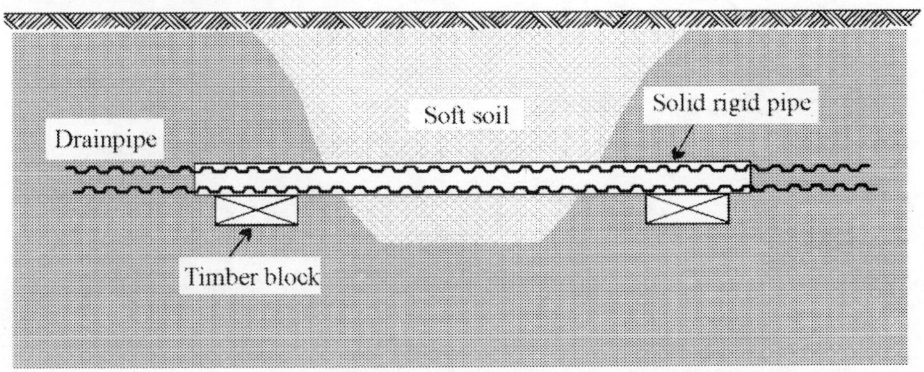

Figure 9.8. Drain Bridge

Drainpipes can be connected to or slid into a rigid, reinforced concrete, plastic or coated steel pipe where they have to cross a road, a waterway, a gutter, unstable soil, a row of trees to prevent roots from growing into the pipes, or other obstacles.

9.10. GRADIENT REDUCERS

A gradient reducer may be required in sloping lands to reduce excessive flow velocities in drain pipes and prevent erosion and subsequent water movement through channels formed outside the pipe. They can be made of concrete or plastic pipes, or of masonry or concrete). They are in fact blind junction boxes of great height with the entering pipe near the top and the leaving pipe near the bottom of the box.

CHAPTER – 10

RECLAMATION OF SALT AFFECTED SOILS

For agricultural purposes, the soils that contain excessive concentrations of soluble salts or exchangeable sodium or both are referred to as problem soils. These salts seriously hamper crop growth and yields.

Irrigation is one of the most important inputs to agricultural production system specially in arid and semi-arid areas. However, owing to changes in natural hydrologic balance after introduction of irrigation, most of these areas experience rise in water table. This rise in water table leads to water-logging and consequently salinisation which, in turn, seriously hampers crop production. The reclamation and management of salt-affected soils has become important and urgent due to ever increasing pressure on land resources to meet food requirements of increasing population and also to mitigate the perils of irrigated agriculture in many irrigation project command areas. In addition, since salt-affected soils commonly occur in productive irrigated lands, their reclamation and strategies to prevent further proliferation is of paramount importance.

Salt-affected soils cover an area of about 8.57 million ha in India, almost half of it lies in canal command areas. Five states, namely; Haryana, Punjab, Rajasthan, Gujarat and Andhra Pradesh accounts for 48 per cent of total salt affected soils of the country. Based on pH, exchangeable sodium, quantum and nature of salts, these soils have been classified in three groups. They are alkali, saline, and saline alkali.

The processes and practices involved in bringing saline and alkali soils into productive condition are known as reclamation measures. The efficiency of any reclamation measure depends on the proper diagnosis of the problem soils. Hence, before any reclamation procedure is applied, it is essential to determine (i) the nature of the soil, as saline, alkali or saline-alkali, (ii) degree of salinity or alkalinity in the soil profile, (iii) quality of water available for leaching out the salts and the reaction products, (iv) drainage characteristics of the soil, (v) topography of the land, and (vi) presence of any hard pan of lime or clay in the soil strata.

10.1. SALT AFFECTED SOILS

The salt affected soils have been classified into three categories; alkaline, saline and saline-alkali; which require distinct management and reclamation procedures.

Soluble salts increase salt content of soil solution which causes harm to plants. When the soluble constituents are mainly sodium salts, the exchangeable sodium with increasing degree of saturation become permanent in nature. Saline soil refers to a soil that contain sufficient quantities of soluble salts capable of impairing soil productivity. In case of alkali soils, productivity is affected by exchangeable sodium. Thus, alkali soils may or may not contain excess soluble salts. The soils in excess of both the soluble salts and the exchangeable sodium content are referred as saline-alkali soils.

Alkali Soils

Alkali soils, also known as sodic soils and also referred as non-saline alkali soil, are characterized by an exchangeable sodium per centage (ESP) more than 15 and pH of saturated soil paste more than 8.5. However, the pH value may be lower if ESP does not greatly exceed 15 or alkaline carbonates are absent in soil. In such situations, pH value can be 8.2. But, in general, the pH value for these soils ranges between 8.5 and 10. The electrical conductivity (EC) of the saturated paste of these soils is variable but normally less than 4 dS/m (decisiemens/metre) or 4 mmhos/cm. These soils are dominated by the presence of Na_2CO_3 and $NaHCO_3$ and contain 2 to 4 per cent amorphous $CaCO_3$ on the surface soil and also a hard Kankar pan made of $Ca\ CO_3$ in the sub-soil. These soils are highly dispersed and have poor air and water permeability affecting plant growth.

Alkali soils occupy extensive areas in the otherwise most productive land in the country, specially India-Gangetic plains, resulting into huge loss of potential crop production. In addition, these soils causes havoc in other forms also, like they are poorly permeable and absorb little or no rain water resulting into the loses of excellent quality rain water. Thus, even small and medium intensity monsoon rains, may accentuate flood problem causing damage to growing crops and property. Large amounts of runoff also carry huge quantities of surface soil causing silting of reservoirs and loss of plant nutrients. These losses combined with other environmental impact on society warrants suitable measures for proper management of alkali soils. The alkali soils of India fall under three classes. They are,

(a) Alkali soils of Indo-Gangetic plain with sweet ground water,

(b) Alkali soil of Indo-Gangetic plain with sodic ground water, and

(c) Alkali soils of Deccan Plateau with sodic or saline ground water.

Reclamation of these soils is achieved by replacing exchangeable sodium. Gypsum is a suitable amendment in such soils. The other acid or acid type amendments may cause excessive acidity unless limestone is also applied.

Saline Soils

Saline soils are characterised by pH less than 8.5, ESP less than 15 and EC generally more than 4 dS/m. Excess amounts of soluble salts are mainly chloride and sulphates of Na, Ca and Mg limit optimal crop growth due to direct toxic effects of salt and increased osmotic stress which physiologically inhibits the availability of soil water. The soluble sodium content exceeds those of calcium plus magnesium in these soils.

In India, six classes of saline soils are found namely;

(a) Saline soils of semi-arid and arid regions with highly saline ground water,

(b) Saline soils of sub-humid region with moderately saline ground water,

(c) Saline soils of sub-humid delta region,

(d) Saline soils of semi-arid delta region,

(e) Saline acid sulphate soils of humid regions, and

(f) Saline marsh of arid region.

Reclamation of these soils is achieved mainly be leaching, thus providing adequate drainage facilities as well. As stated earlier, saline soils contain neutral salts e.g. chlorides and sulphates of sodium, magnesium and calcium whereas alkali soils contain salts capable of alkaline hydrolysis e.g. sodium carbonate, bicarbonate and silicate.

Saline-Alkali Soils

These soils are characterised by EC of saturation extract greater than 4 dS/m, ESP more than 15. The pH value of saturation extract is about 8.5 when excess salts are present. The pH may exceed 8.5 after leaching of salts, thus soil becoming strongly alkaline. After leaching of salts, the soil becomes unfavourable for entry and movement of water and also for tillage. Sometimes saline-alkali soils contain gypsum. Once the soils are properly leached, calcium dissolves and replaces exchangeable sodium concurrently with removal of excess salts.

Table 10.1. Classification of Salt Affected soils

Soil Type	pH	ESP%	EC in dS/m
(1) Alkali	8.5 -10.0	>15	<4
(2) Saline	<8.5	<15	>4
(3) Saline-alkali	about 8.5	> 15	>4

10.2. SOURCES OF SALINITY

Over irrigation with poor water and poor water management are the chief causes of water-logging and salt-accumulation on surface. Development of soil salinity is a challenge to the permanence of irrigated agriculture. It is very important to understand the causes of salinity and alkalinity of soil to enable you to suggest precautions to be followed in affected areas.

Sources of Soluble Salts

The soluble salts that occur in soils consist of the cations of sodium, calcium, and magnesium, and the anions of chloride and sulfate. Constituents that ordinarily occur only in minor amounts are the cation of potassium and the anions of bicarbonate, carbonate, and nitrate. The direct source of all the salt constituents are the primary minerals found in soils and the exposed rocks of the earth's crust. During the process of chemical weathering, which involves hydrolysis, hydration, solution, oxidation, and carbonation, these constituents are gradually released and made soluble.

Bicarbonate ions form as a result of the solution of carbon dioxide in water. The carbon dioxide may be of atmospheric or biological origin. Water containing carbon dioxide is a particularly active chemical weathering agent that releases appreciable quantities of the cation constituents as the bicarbonates. Carbonate and bicarbonate ions are interrelated and are functions of pH. Appreciable amounts of carbonate ions can be present only at pH values of 9.5 or higher.

Weathering of primary minerals is the indirect source of nearly all soluble salts. Saline soils usually occur in areas that receive salts form other locations, and water is the primary carrier. The ocean may be the source of salts as in soils where the parent material consists of marine deposits that were laid down during earlier geologic periods and have since been uplifted. The ocean is also the source of the salts in low-lying soils along the sea-coasts. Sometimes salt is moved inland through the transportation of spray by winds. The direct source of salts is surface and ground waters. All of these waters contain dissolved salts, the concentration depending upon the salt content of the soil and geologic materials with which the water has been in contact. Waters act as sources of salts when used for irrigation. They may also add salts to soils under natural conditions, as when they flood low-lying land or when ground water rises close to the soil surface. We shall discuss it in next section.

Salinisation of Soils

Saline soils occur for the most part in regions of arid or semi-arid climate. Under humid conditions the soluble salts originally present in soil materials and those formed by the weathering of minerals generally are carried downward into the ground water and are transported ultimately by streams to the oceans. Saline soils are, therefore, practically nonexistent in humid regions, except when the soil has been subjected to sea water in river deltas and other law-lying land near the sea. Leaching is usually local in nature, and soluble salts may not be transported far because there is less rainfall available to leach and transport the salts. High evaporation rates of arid climates tend to concentrate the salts in soils and in surface waters.

Restricted drainage usually contributes to the salinisation of soils and may involve the presence of a high grand water table of low permeability of the soil. The high ground water table is often related to topography. Owing to the low rainfall in arid regions, surface drainage ways may be poorly developed. As a consequence, there are drainage basins that have no outlet to permanent streams. The drainage of salt-bearing waters away from the higher lands of the basin may raise the ground water level to the soil surface on the lower lands. This may cause temporary flooding or may form permanent salty lakes. Under such conditions upward movement of saline ground water or evaporation of surface water results in the formation of saline soil.

Low permeability of the soil causes poor drainage by impeding the downward movement of water. Low permeability may be the result of an unfavourable soil texture of structure of the presence of indicated layers. The latter may consist of a clay pan, or a silica hardpan. The previously nonsaline soil becomes saline due to over irrigation coupled with poor drainage conditions. Due to poor drainage, water table rises faster.

10.3. CAUSES OF WATER-LOGGING

The main causes of water-logging are the following:

1. Losses from main and branch canals and distributaries,
2. Seepage from the canal,
3. Faulty on-farm water management,
4. In adequate outlets and Lack of natural drainage,
5. Upheaval in the river bed or in the bed of drainage channel,
6. The cultivation in bed of drainage channels or putting of fishing barriers,
7. Interception of natural drainage channels due to construction of roads, embankment etc,
8. Non-utilization of ground water,
9. Back water flows or inundations in coastal areas during cyclonic storms,
10. Use of saline irrigation water,
11. Deposition of salts on soil surface from high sub-soil water table, and
12. Excessive application of irrigation water in some states where electricity for agricultural purpose is free of cost or highly subsidised.

10.4. SOURCE OF SALTS IN WATER

The weathering of parent material of soil or rocks through processes like hydrolysis, hydration, solution forming, oxidation and carbonation, result in entry of salts in irrigation water. The case of weathering of a mineral depends on the basicity of mineral, degree of linkages of tetrahydrons in the clay mineral-structure and the degree of crystallinity and purity of minerals. Generally weathering of the minerals is governed by one or more causes of water-logging. The possible sources of excess salt in soils are given below.

1. High salt deposits inherited by the soil from the original parent material during soil forming process,
2. Salts contained in the irrigation water applied or in the water lost in conveyance through irrigation distributions system,
3. Salt may come through upward movement (capillary action) of shallow brackish ground water,
4. High sub-soil water table thus poor drainage conditions,
5. Back water flow or intrusion of sea water in coastal area
6. Seepage from canals and farm-irrigation systems due to injudicious use of water, and
7. High cropping intensity and replacement of low water requiring crops by crops of high water demand thus, gradual rise adds in raising water table.

10.5. RECLAMATION OF ALKALI SOILS

As stated, reclamation of alkali soils, basically, requires replacing of excess sodium on the soil exchange complex by calcium through amendments and leaching out of the exchanged sodium from the root zone. For any long lasting reclamation programme, provision of adequate drainage is a key factor. Reclamation procedure must address to socio-economic condition of farmers to offset the cost incurred.

The following are the possible options to reclaim alkali soils.

(a) Surface and Sub-surface Drainage

The drainage of alkali soils is difficult. Studies have proved that both surface and sub-surface drains are ineffective in lowering the ground water table and removing of salts because of extremely poor permeability of alkali soils. However, in some areas in the state like Haryana, provisions of sub-surface drainage and use of amendment have proved successful. Open drainage ditches spaced at suitable distance are also found to be effective.

(b) Vertical Drainage

Vertical drainage system has proved to be an effective measure for controlling ground water table in alkali areas. Vertical drainage system may follow two methods. In one, drainage water is disposed through wells into a porous layer of earth or on open rock formations. Drainage is affected by conducting surface water through vertical wells into water-absorbing stratum. A substratum capable of taking large volume of water rapidly is needed for such drainage system, which is rarely found. But, where conditions are favourable, this method of drainage possesses considerable merit. In other method of vertical drainage a borehole is made through which underlying ground water is extracted by pumping. This lowers the ground water table. In alkali soil, ground water quality is generally good but in some cases it may contain high residual sodium carbonate. Success of a drainage well system depends on presence of a favourable aquifer within 10-20 m with good quality water. Large-scale exploitation of ground water for irrigation purpose in alkali soil areas have proved to be affective in lowering ground water table. Thus, vertical drainage provides an ideal measure for controlling ground water table in alkaline soil areas.

(c) Rain Water Management in Alkali Soils

As explained to you, rain water coming out of even small and medium storms poses a problem in alkali areas, it becomes very important to manage rain water of excellent quality. The following 3-tier system of rainwater management is suggested.

1. Conserve the rainfall in cropped land till such time and to the extent that it will not harm the standing crop.
2. Direct the excess water from various parts of the catchment into the dug out storage ponds of sufficient capacity located at the lowest elevation of the catchment. This stored water can be utilised for irrigation in the adjacent fields during the dry spells of the monsoon season.
3. Allow the remaining excess water to discharge into the shallow surface drains provided for the area.

(d) Amendments Method

The replacement of sodium by calcium through application of chemical amendments like gypsum ($CaSO_4$. $2H_2O$). Gypsum (powdered hydrated Calcium Sulphate - $CaSO_4$, $2H_2O$) is a commonly used chemical amendment to the alkali soil for replacing the adsorbed Sodium by Calcium. The replaced Sodium, in the form of aqueous solution of Sodium Sulphate, is to be removed from the soil profile through drainage so that it may not again get attached to the soil when Calcium is eventually depleted and the Sodium ion concentration becomes more than the Calcium ion concentration in the soil. A few other amendments like pyrites, calcium carbonate, calcium chloride, phosphogypsum etc. have also been exploited. Table presents equivalent quantities of different amendments needed in reference to chemically pure gypsum (Ca SO_4 $2H_2O$) or elemental sulphur. Being cheapest and abundantly available, gypsum is the most commonly used amendment for soil reclamation. Studies conducted at Central Soil Salinity Research Institute (CSSRI), Kamal (Haryana) and other places in the country have culminated, in the development of a gypsum based viable technology and associated agronomic and cultural practices for reclamation of these soils. The improvement of soil tilth is slow in alkali soils. Excessive exchangeable sodium can be replaced by any cheap source of calcium and magnesium salts such as calcium chloride, magnesium chloride and gypsum. Soluble calcium and magnesium salts are suitable for the reclamation of alkali soils.

Table 10.2. Chemical Properties of Various Amendments for Reclaiming Alkali

Amendment	Chemical Composition	Physical Dissension	Solubility in Cold Water	Amount* Equivalent to 1 kg of Gypsum (kg)	Amount* Equivalent to 1 kg of Sulphur (kg)
Gypsum	$CaSO_4.2H_2O$	White mineral	2.4	1.0	5.38
Sulphur	S	Yellow element	Insoluble	0.19	1.00
Sulphuric acid	H_2SO_4	Corrosive liquid	Very high	0.57	3.06
Calcium carbonate	$CaCQ_3$	White mineral	0.014	0.58	3.13
Calcium chloride	$CaCl_2.2H_20$	White salt	Very high	0.85	4.59
Ferrous sulphate	$FeSO_4.7H_20$	Blue green salt	1.56	1.61	8.69
Ferric sulphate	$Fe2(SO_4)39H_20$	Yellow-brown salt	Very high	1.09	5.85
Aluminium sulphate	$A1_2(S0_4)_3.18\ H_2O$	White mineral salt	Very nigh	1.29	6.94
Pyrites (30% S)	FeS_2	Yellow-black mineral	0.005	0.63	1.87

* Amount based on chemically pure

The choice of the ameliorants depends on the soil pH, the amount of calcium or magnesium carbonate present, cost of the ameliorants, speed of the reaction desired, soil condition, quality of water available and the cost-benefit ratio of reclamation. Gypsum in a powdered form is usually broadcast on the field and then mixed with the top soil layer using a suitable implement (e.g. a disc plough). Following broadcast and mixing, adequate water is added to the field to dissolve the gypsum.

Gypsum Requirement (GR)

The quantity of gypsum required for the improvement of alkali soils depends on the quantity of the exchangeable sodium that must be replaced and the depth to which soil improvement is desired. For soils which contain soluble sodium carbonate, as most alkali soils of the Indo-Gangetic plains do, the estimation of gypsum requirement is done to neutralise the soluble carbonates and to replace exchangeable sodium. Another fact to be taken into account is that only a small fraction of the gypsum applied at surface reacts with soluble carbonates to precipitate $CaCC_3$ as $CaSO_4$, during the leaching process and most soluble carbonates are leached practically unreacted. In the light of the above, gypsum requirement may be calculated using following equation:

$$GR = 8.6\, \rho_b\, D_s\, CEC\, (E_{Nai} - E_{naf}) \times n$$

where,

GR = gypsum requirement (tonnes/hectare),

ρ_b = soil bulk density (tonnes/m),

D_s = depth of soil to be amended (cm),

CEC - cation Exchange capacity (mmol/kg of dry soil),

E_{Nai} = initial equivalent fractions of exchangeable Na,

E_{naf} = final equivalent fraction of exchangeable Na, and

n = Na-Ca exchange inefficiency factor

Na-Ca exchange inefficiency factor (n) is a measure of gypsum loss and its value is usually taken as 1.2. Considering that lm^3 of soil weights approximately 1.5 tonnes, amount of gypsum required to replace exchangeable sodium is given in Table below.

Table 10.3. Gypsum Requirement and Exchangeable Sodium

Exchangeable Sodium (meq per 100 g of soil)	Gypsum Requirement (tonnes/ ha-m)
1	13.9
2	27.8
3	41.8
4	55.7
5	69.6
6	83.5
7	97.4
8	111.3
9	125.1
10	139.2

Depending upon soil texture and initial pH, 4 to 16 tonnes/ha of gypsum is required to reclaim the upper 15 cm soil for raising rice and wheat crop but in some areas this quantity may go upto 40 tonnes/ha. A convenient method using a graphical approximation for determination of GR for alkali soils has been developed and presented in Figure 10.1.

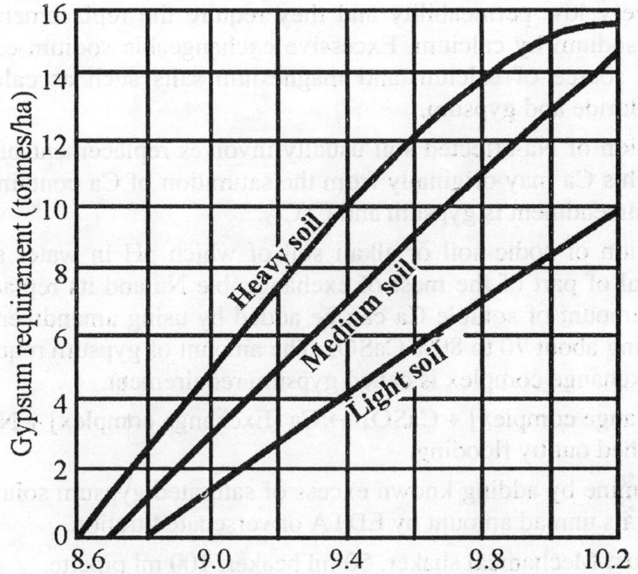

Figure 10.1. Relationship between Gypsum requirement and Soil pH

An abridged sample table of gypsum requirement for selected values of milliequivalents of Sodium replacement and for three dry densities of field soil is given in Table 10.4.

Table 10.4. Gypsum Requirement Using Pure Gypsum, One-to-one Replacement of Sodium by Calcium and Complete Dissolution and Mixing of Gypsum

Sodium Replacement meq/100 g of Soil Dry Density	**Gypsum Requirement (t/ha) for Reclaiming15 cm Field Soil at the Following Average**		
	1.2g/cm^3	**1.4 g/cm^3**	**1.6 g/cm^3**
2	3.1	3.6	4.1
4	6.2	7.2	8.3
6	9.3	10.9	12.4
8	12.4	14.5	16.5
10	15.5	18.1	20.7

1. The followings points are important for better and sustained results Reclamation can be sustained and improved by continuous cultivation, with rice as one of the crops in the cropping sequence.
2. The amount of gypsum can be reduced by application of farm yard manure (FYM) at the rate of 10 tonnes per hectare and by growing salt tolerant varieties.

Determination of Gypsum Requirement for Land Reclamation

The reclamation of alkali soils is more difficult than saline soils because alkali soils have a very low permeability and they require the replacement of excessive exchangeable sodium by calcium. Excessive exchangeable sodium can be replaced by any cheap source of calcium and magnesium salts such as calcium chloride, magnesium chloride and gypsum.

Reclamation of Na-affected soil usually involves replacement of exchangeable Na with Ca. This Ca may originally from the saturation of Ca containing mineral in the soil. Such amendment is gypsum and $CaCl_2$.

Reclamation of sodic soil or alkali soil of which pH in water suspension 8.5 require removal of part of the most of exchangeable Na and its replacement by Ca. The required amount of soluble Ca can be added by using amendment like gypsum which containing about 70 to 80% $CaSO_4$. The amount of gypsum requires to replace Na ion from exchange complex is called gypsum requirement.

Na [Exchange complex] + $CaSO_4 \rightarrow$ Ca [Exchange complex] + Na, SO_4 leached Na_2SO_4 is leached out by flooding.

GR determine by adding known excess of saturated gypsum solution to the soil and estimating its unread amount by EDTA or versenate titration.

Apparatus: Mechanical shaker, 50 ml beaker, 100 ml pipette.

Reagents

1. Saturated gypsum solution

Shake 5 g $CaSO_4$ with 1 liter of distilled water for 10 minutes on mechanical shaker and filtered (28 to 30 me Ca/l).

2. Buffer solution for 10 pH

Dissolve 67.5 g NH_4Cl in 570 ml NH_4OH and dilute to 1 liter.

3. Erichrome Black- T (EBT)

Dissolve 0.5 g EBT and 4.5 g hydroxylamine hydrochloride in 100 ml of 95 % ethanol.

4. Standard $CaCl_2$ solution 0.01 N

Dissolve 0.5 g pure $CaCO_3$ in 10 ml of 3 N HCI and dilute to 1 liter.

5. EDTA solution 0.01 N

Dissolve 2 g EDT A and 0.05 g $MgCl_2$ in 900 ml water and standardize with 0.01 N $CaCl_2$ solution.

Procedure

Take 5 g air dried soil into 250 ml conical flask and add 100 ml saturated gypsum solution with volumetric pipette. Stop the bottle. Shake for 5 minutes on mechanical shaker or 30 minutes on rotary shaker and filter it. Pipette out 5 ml of soil extract into 100 ml conical flask and dilute to 25 ml distilled water. Add 0.5 ml buffer solution, 3-4 drops of EBT indicator and titrate with 0.01 N standard EDTA solution until the colour change from wine red to blue. Titrate similarly 5 ml of saturated gypsum solution. In the modified method of GR, Abrol *et al.* (1975) suggested washing with alcohol before adding gypsum saturation solution. Remaining procedure is same.

Observations and calculation

Table 10.5. Observation table for Gypsum requirement

Sample No.	GR as per Schoonover method		GR as per modified Schoonover method	
	Reading	**Difference**	**Reading**	**Difference**
Blank	14.65	-	14.65	-
1	14.05	0.60	13.60	1.05
2	14.00	0.65	13.70	0.95
3	11.65	3.00	11.45	3.20
4	14.30	0.35	14.35	0.30

- 1 N 1 liter EDTA = 86 g $CaSO_4.2H_2O$ (gypsum)
- N 1 ml EDTA = 0.01 me = 0.86 mg gypsum
- % GR = Difference x (0.86/1000) x (100/5) x (100/5)
- ppm GR = Difference x (0.86/1000) x (100/5) x (100/5) x 10000
- kg/ha GR =Difference x (0.86/1000) x (100/5) x(100/5) x 10000 x 2.24
- t/ha GR = Difference x 7.7
- GR meq/100 g = (Ca cone. of added gypsum solution in meq/l - Ca + Mg cone. of filtrate in meq/l) x 2

Efficiency of gypsum application

The efficiency of a gypsum application depends on various factors:

- Not all of the gypsum will be used to replace sodium; some of it will replace other cations like Mg and K;
- The distribution of gypsum will be irregular (i.e. there is a non-uniform spreading);
- Part of the gypsum will be transported with the percolating water into the subsoil, and will not replace sodium in the topsoil. This phenomenon may be more pronounced in soils rich in soluble sodium chloride. This salt enhances the solubility of gypsum;
- The fineness of the gypsum;
- The type of soluble salts present in the soil solution.

Gypsum Application Methods

Field trials and laboratory investigations have shown that mixing required quantities of gypsum at the desired reclamation depth has proved advantageous than mixing in deeper depths. Mixing at greater depth reduces its reclamation efficiency. Also, shallow mixing reduces surface crusting and infiltration problems. Better results are obtained if gypsum is reduced to small size before applying which can pass through 2 mm sieve.

(e) Biological Methods

Amelioration in the alkali soils may also be achieved through biological activities of the roots, specially of rice, grasses like Karnal grass *(Leptochloa fusca)* and trees like *Prosopus fitlifloran*, *Acacia nilotica* and *Sesbania aegyptica.* Such forms of biological reclamation can be taken up at community basis on Panchayat land to improve forest cover and environment. It can provide fuel wood and animal fodder as well as can generate additional employment.

Green manuring

Highly dispersed alkali soils are devoid of soil organic matter and green manuring with dhaincha (Serbania aculeatd) in summer is helpful in reclaiming such soils. Green manuring adds soil organic matter and increases soil fertility. The dhaincha is tolerant to salts and waterlogging conditions, its root system is extensive, it opens compact subsoil and improves soil permeability, its rate of decomposition is fast, and its cell sap is acidic and is capable of neutralizing the alkalinity of the soil.

Crops in Alkali Soils

If resources for reclaiming soils by different methods described above are not available, growing of salt tolerant crops/varieties have also been found to reduce soil pH and ESP. Table 9.4 gives relative crop tolerance to alkalinity and help's in choosing appropriate crops. Rice is very important in any cropping sequence recommended for adoption during reclamation of alkali soils. It has high tolerance to soil sodicity and submerged soil conditions. Growing rice enhances hydraulic conductivity due to physical action of plant roots by facilitating water movement and to the removal of entrapped air from the larger conducting pores. The FYM also improves hydraulic conductivity of soils through chemical action of CO_2 on calcite, naturally present in alkali soils.

Table 10.6. Crop Tolerance to Alkalinity of Soil

ESP Range	Crops
10-15	Safflower, peas, lentil, pigeon pea.
16-20	Bengal gram/soybean
20-25	Groundnut, cowpea. onion, peatlmillet.
25-30	Linseed, garlic, guar
30-50	Raya, wheat sunflower
50-60	Barley, sesbani
60-70	Rice

10.6. RECLAMATION OF SALINE SOIL

Management of water-logged saline soils basically involves lowering of water table below the root zone of crop and leaching of excess salts. In general, water with high salt content should not be used for irrigating saline soils or the soils that have low infiltration and drainage rates. Use of ground water in such areas for irrigation is found to be useful in these soils if the aquifer is unconfined or partially confined because this helps in keeping water table below root zone. This indirectly also reduces wasteful over use of water as the pumping costs are involved.

To reduce the risk of soil is becoming saline, over use of irrigation water be curtailed at any cost. Canals be lined so as to reduce seepage and avoid water-logging in poorly drained soils and thus salinity. The irrigation methods which encourage wastage of water be discouraged in areas where drainage is a problem. However, in order to reduce the salinity of existing saline soils, flooding method is favourable as it would help in leaching the salts below the root zone. If water table is shallow or natural sub-surface drainage is restricted, the percolated salt solutions are to be removed by installing a sub-surface drainage system. Even so, the permanent reclamation may be difficult as, the hot and dry weather prevailing over several months, particularly in monsoon climate as in India, resalinizes the soil profile.

Leaching is the only practical way to remove excess salts from the soil which is achieved by displacing saline soil solution with good quality water or with water of lower salt concentration. The reclamation depends on the efficiency of the removal of salts from the upper to the lower layer in case drainage is not a problem. The following methods may be employed, in combination, if necessary, to ameliorate the saline soil condition.

Leaching

Leaching is the simplest procedure, in which the field is flooded after making ridges or levees at the boundaries. The movement of water through soil profile will carry dissolved salts into the lower strata below the root zone. These salts are drained away, if proper sub-surface drainage is provided. Without leaching, salts accumulate in direct proportion to the salt content of irrigation water and depth of water applied. The quality of water needed for leaching depends mainly on the severity of the salinity problem and the depth and degree of reclamation desired. The lower the salt content in the irrigation water, the greater is the efficiency in reclaiming saline soils. In case there is no leaching of salts in soil, even reasonably good quality of irrigation water can turn a soil into a saline soil.

In the coastal region, there may be considerable upward flux of saline groundwater from below and that would also require long-term functioning of the sub-surface drainage system to continuously remove salts through leaching. A well-designed and executed sub-surface drainage system has a long life and the leaching requirement is automatically taken care as long as fresh water is used for irrigation and if required, additional fresh water is used for leaching. The quantity of water needed for leaching depends mainly on the severity of the salinity problem and the depth and degree of the reclamation desired.

The relationship between electrical conductivity value of irrigation water and saturation extract of soil can be given by the following equation:

$$\frac{Diw}{Ds} = \left(\frac{\rho s}{\rho w}\right)\left(\frac{SP}{100}\right)\left(\frac{\Delta ECe}{ECiw}\right)$$

where,

Diw = depth of irrigation water,

Ds = depth of soil,

ρw = density of water,

ρs = density of soil,

SP = saturation per centage (moisture content of soil at saturation)

ΔECe = change in electrical conductivity of saturation extract of a soil depth (Ds), and

ECiw = electrical conductivity of irrigation water.

Leaching requirement

The leaching requirement may be defined as the fraction of the irrigation water that must be leached through the root zone to control soil salinity at any specified level. This concept has greatest usefulness when applied to steady-state water-flow rates or the total depths of water used for irrigation and leaching over a long period of time. In general, irrigating with wastewater or saline irrigation water presents specific risks. Among these risks are soil salinization and salinity hazards due to salt accumulation in the root zone, especially in areas affected by under-irrigation. Although saline soil can produce acceptable yields, excessively saline irrigation water leads to reduced water available for plant use, which in turn can result in lower stem diameter and subsequently, lower fruit yield. In order to compensate for the salt accumulation, irrigation with highly saline water re- quires larger and more frequent applications than irrigation with good quality water. The classic explanation of leaching effectiveness by Hoffman (1986) describes vertical leaching in soils. Leaching with closely spaced furrows, sprinkler irrigation on row crops, and border strip and basin irrigation (if on non-cracking soils, and ignoring preferential flow) is approximately vertical. Hoffman showed that the effectiveness of vertical leaching depends on the per centage of deep percolation, the soil, and the rate at which water infiltrates into the soil. In particular, he noted that intermittent ponding was more effective in salt leaching than was continuous ponding. The most effective and efficient reclamation leaching practices for tree crops irrigated with drip appear to include:

1. Apply leaching water only to the areas with salt accumulation, typically along the tree row with drip lines.
2. Use low application rates for maximum effectiveness of salt removal.
3. Consider the point of diminishing effectiveness for reclamation leaching: quantities of leaching water greater than 0.8 equivalent depth may result in insignificant salt reduction (for a typical silt loam soil using intermittent leaching).

4. Use intermittent applications of leaching water, which minimize the effects of bypass flow.

Leaching is the removal of soluble salts from soil by the passage of water through it. It is the fraction of water entering the soil that must pass through the root zone in order to prevent soil salinity from exceeding a specified limit. It is the amount of water to be applied, in addition to the water requirement of crops, to leach the salts through the root zone. If salt is accumulated in the soil or on the surface because of high water table or irrigation with saline water, it may be washed downward and removed by the process of leaching. Leaching requirement is usually expressed under steady state or long time average conditions. The leaching requirement in subsurface drainage design may be obtained through the analysis of climatological data and the values of salt concentration of the soil and irrigation water. It may be computed from the following relationship

$$LR - (E\text{-}P)\, C_i / F(C_{fc}\text{-}C_i)$$

Where

LR = Leaching requirement, mm

E = Evapotranspiration, mm

P = Effective rainfall, mm

C_i = Concentration of salt in irrigation water, me/l

Salt concentrations are computed using the general relationship:

Concentration in me/l = 12 x electrical conductivity in dS/m and salt concentration at field capacity = 2 x concentration of 1 : 2 saturation extract.

C_{fc} = Concentration of salt in the soil at field capacity, me/l (me/t = milliequivalent per litre, Equivalent, or equivalent weight is the weight in grams of an ion or compound that combines with or replaces one gram of hydrogen. Milliequivalent means one-thousandth of an equivalent. Milliequivalent per litre is the milliequivalent of an ion or a compound in a litre of the solution.)

F = Leaching efficiency, fraction (Leaching efficiency of 0.4 is assumed as the soil is clay (for light soils it is 0.6 to 0.8)

The average rainfall may be taken as effective rainfall for the flat topography area with bunding on all sides.

The leaching requirement depends upon the salt concentration of the irrigation water and upon the maximum concentration permissible in the soil solution. The maximum concentration, except for salt crusts formed by surface evaporation, will occur at the bottom of the root zone and will be the same as the concentration of the drainage water from a soil where irrigation water is applied with areal uniformity and with no excess leaching. Increase of the concentration of salts from the value existing in the irrigation water to the value occurring in the drainage water is related directly to consumptive use. This mostly depends upon water extracted from the soil by roots and on the salt tolerance of the crop.

The maximum concentration of the soil solution should be kept below 4 mmhos/cm for sensitive crops. Tolerant crops like beets, alfalfa, and cotton may give good yields up to 8 mmhos/cm, while a very tolerant crop like barley may give good yields at values of 12 mmhos/cm or higher.

Assume uniform areal application of irrigation water; no rainfall; no removal of salt in the harvested crop; and no precipitation of soluble constituents in the soil. Let calculation be based on steady-state-flow rates. Thus, the leaching requirement (LR) is simply the ratio of the equivalent depth of the drainage water to the depth of irrigation water (Ddw/DiW) and may be expressed as a fraction or as per cent. Under the foregoing assumed conditions, this ratio is equal to the inverse ratio of the corresponding electrical conductivities, that is:

$$LR = \frac{Ddw}{Diw} = \left(\frac{EC_{iw}}{EC_{dw}}\right)$$

Leaching requirement for a particular crop can be worked out by following formula:

$$LR = \left(\frac{EC_{iw}}{5\,(EC_c) - EC_{iw}}\right)$$

where,

LR = Leaching requirement as a portion of the infiltrated water,

ECiw = EC of irrigation water, and

ECc = Average soil salinity tolerated by specific crop

Table 10.7. Average soil salinity tolerated by specific crop

Crop	EC_C (mmhos/cm)
Barley	8.0
Cotton	7.7
Sorghum	4.0
Soyabean	5.0
Wheat	6.0
Tomato	2.5

Rao *et al.* (1990) used the subsurface drainage system for the reclamation of waterlogged saline alluvial soils in India. The quantity of water needed for leaching depends mainly on the severity of the salinity problem and the depth and degree of the reclamation desired. The lower the salt content in the irrigation water, the greater is the efficiency in reclaiming saline soils. In light soils higher rate is favourable but in heavy soils intermittent flushing is better, as the water gets adequate time to percolate slowly through the small soil pores. This excess water required to meet the leaching needs, is generally expressed as the per centage of the total irrigation water applied to the soil (field) to meet the consumptive use as well as the leaching needs. The leaching requirement (LR) is the difference between the deep percolation from the root zone and the capillary flow from groundwater to the root zone, both expressed in units of depth of water over a certain period (e.g., mm/day, mm/month and so on. Under irrigation or rainfall, the outflow from the root zone is always greater than the inflow to the root zone due to upward capillary movement of water from deeper layers. Hence, LR is a positive quantity.

Estimation of leaching requirement

Leaching Requirement

$$LR = \frac{\text{Depth of water drained out per unit area}}{\text{Depth of irrigation water applied per unit area}}$$

$$LR = \frac{D_d}{D_i}$$

Leaching efficiency

The efficiency of the reclamation per unit volume of water will be maximum with good water. The lower the salt content in the irrigation water, the greater is the efficiency in reclaiming saline soils. In practice, where water is inadequate or the soils are slowly permeable, it is desirable to obtain partial reclamation and then a more salt tolerant crop like barley could be grown. Leaching may be continued during the cropping period also. The effectiveness of leaching varies from soil to soil and depends largely on the quantity of water that passes through the soil. The rate of removal of salt is fast in the beginning but later on the process is slowed down. The efficiency of leaching is also influenced by varying the water flow velocities within a soil. In light soils higher rate is favorable but in heavy soils intermittent flushing is better, as the water gets adequate time to percolate slowly through the small soil pores.

The leaching efficiency is low (0.2 to 0.5) for fine textured soils and is relatively higher (> 0.5) for coarser soils. The leaching operation becomes more efficient if the calculated leaching requirement is given in split applications rather than dumping the entire water on the soil at one time. For an effective leaching, the field should be bunded in small units and levelled such that there is no runoff loss and the water spread is of uniform depth on each land unit to be leached.

Factors affecting Leaching efficiency

The factors on which the efficiency of leaching depends are: (a) the amount of water applied, (b) uniformity of water distribution, and (c) the adequacy of drainage. In case of salt tolerant crops it is sufficient to leach once or twice in a growing season. But in less tolerant crops, it is advisable to apply irrigation water several times for leaching during the crop period. The best practice would be to follow the salinity trends by making regular checks, in the cultivated fields irrigated with saline waters. Since nutrients are also moved down along with water and salts, it is desirable to apply fertilizers after leaching and make up the losses of the nutrients, if any, in leaching.

Saline sodic soils often show a good soil structure at first. After drainage and an initial leaching, when a large portion of the salts has been removed from the topsoil, the breakdown of soil structure becomes evident: the infiltration rate decreases, the process of leaching is slowed down, and the reclamation period is prolonged.

The depth of water required and the time needed for reclamation will vary from place to place, depending on: Soil properties: soil texture, soil structure, soil pore geometry, cracking phenomena, clay mineralogy, and physico-chemical behaviour of clays; Initial salt content and the chemical composition of the soluble salts; Salt content and the chemical composition of the water used for leaching (rain or irrigation); Leaching technique and the scope for planting a salt-tolerant crop during reclamation.

Drainage

Provision of adequate surface and sub-surface drainage to lower water table depth and to facilitate leaching of salts is of fundamental importance to reclamation and management of saline soils. In most irrigation project in India, appropriate drainage measures were either not considered at the design and planning stage or were not provided when problems were faced. In certain case ill-designed and mismanaged drainage system caused serious problems.

Drainage Requirement

A drainage system must be adequate to remove from the soil the equivalent depth of water that must be passed through the root zone so as to maintain a favourable salt balance. The drainage requirement can be found out by equation:

$$Ddw = \left(\frac{EC_{iw}}{EC_{dw} - EC_{iw}} \right) D_{cw}$$

Salt tolerance of crop is taken in to account in selecting permissible value of ECdw- As stated earlier, the value of ECdw is taken as two times of maximum permissible value of EC of soil extract that can be tolerated by a crop.

Vertical Drainage

In vertical drainage, the underground water may be removed by means of a bore hole/well either under artesian pressure or through pumping in saline conditions also. Lowering of water table through vertical drainage would prevent the concentration of soluble salts at the soil surface or in the root zone. Suitability conditions for installing a vertical drainage system are listed below:

1. Where a soil stratum of low permeability and of considerable thickness overlies a stratum of high permeability.
2. When it becomes necessary to lower the water table to a considerable depth, specially in saline areas where ground water table is shallow and of poor quality. Thus even with capillary rise, level of the ground water may not reach the root zone.
3. Saline areas requiring low water table depth, with more voids for facilitating the fresh quality rain water recharge and storage.
4. Highly saline soil requiring heavy amount of leaching water to be passed through the affected soil for reclamation purpose.

The vertical drainage have shown good performance in some parts of the country like northern part of Haryana, eastern Punjab, western U.P. This became possible due to presence of suitable aquifer within 12 to 20 m to permit pumping of ground water on a sustained basis and favourable quality of ground water suitable for irrigation.

Bio-drainage

The bio-drainage the lowering the water table through bio-energy of the trees like Eucalyptus. It has been observed that planting of a 100 m or more wide belt along the both sides of canals with high water demanding trees like Eucalyptus, Bopulus and Jamun controls the water-logging caused by seepage. A techno-economically viable sub-surface drainage for safe disposal of saline drainage effluent has been standardised for arid and semi-arid regions. The most practical step for restraining the rise of water table and development of secondary salinisation is to reduce water allocation and improve on farm irrigation management. This should be aimed at reducing volume and salt-load of drainage effluent. For this, the right combination of agronomic and irrigation practices e.g. crop rotation, crop diversification, Argo-forestry, and micro irrigation systems. Conjunctive use of drainage water and fresh water is recommended specially in salinity tolerant crops like barley, wheat and mustard. These crops can tolerate salinity upto 10 ds/m. But these soils need a regular application of nitrogen and phosphorus to maintain crop yield. Generally no deficiency of micro-nutrient has been observed in these soils. Reclamation of salt affected vertisols is more difficult than alluvial soils due to its swelling-shrinking nature, poor internal hydraulic conductivity, susceptibility to sodification and difficulty in laying drainage structure.

10.7. RECLAMATION OF SALINE - ALKALI SOILS

When the soil is saline-alkali, it is necessary to leach out the excessive salts upto an EC value of 6 to 8 ds/m before any attempt is made to replace excessive exchangeable sodium by gypsum. However, some excess salts can be leached by flooding the field after gypsum application, for removing the released sodium from the soil exchange phase. Maintenance of an adequate drainage is essential for the successful reclamation of saline; alkali and saline-alkali soils, as leaching out of the salts or the removal of the reaction products is a must for the success of reclamation. In case the subsoil has lime or hard clay pan, it should be broken by deep ploughing or sub-soiling. Different approaches like surface drainage, sub-surface and vertical drainage, and bio-drainage, either alone or in combination, may offer the best solution depending upon economic, climatic, social as well as edaphic and hydro-geological situations.

Exercises

1. Calculate the monthly leaching requirement of the area where Subsurface drainage is practiced in an acid sulphate clay soil for the controlled leaching of toxic salts. The area has a flat topography and rice is cultivated in bunded fields. Average salt concentration of irrigation water used is 0.25 dS/m. The salt concentration of the 1: 2 saturation extract prepared from the soil is 1.27 dS/m. The above readings were taken during month and the average rainfall of the month is 88 mm. Average evapo-transpiration for the same period was 4.8 mm per day.
2. An irrigation water of 1 mmho/cm electrical conductivity is applied to a soil depth of 30 cm with initial soil electrical conductivity at 2 mmho/cm and 40

per cent moisture content at saturation. After complete evaporation of irrigation water, the saturation extract of soil was found to have 6 mmho/cm electrical conductivity. Calculate the depth of irrigation water applied if density of soil is 1.3 g/cc and no leaching of salts has taken place.

3. Calculate the volume of water required for leaching 5 ha of saline land having electrical conduction 12 dS/m, if the crop grown can tolerate electrical conductivity of 8 dS/m as a maximum. The effective root zone depth to be improved is 1.5 m and water available for irrigation gas an electrical conductivity of 2 ds/m. Take saturation per centage as 40 and specific gravity of soil as 1.6.
4. Alkali soils, also known as sodic soils are characterised by ESP >15, pH between 8.5 and 10 and electrical conductivity normally less than 4 ds/m.
5. When soil possesses characteristics of both alkali and saline soils, it is called as saline-alkali soil. These soils have (ESP) more than 15 and EC > 4 ds/m.
6. In reclaiming saline-alkali soils, it is essential to leach out excess salts upto an EC value of 6 to 8 mmhos/cm followed by removal of excess exchangeable sodium by gypsum.
7. Estimate the gypsum requirement to lower down SAR up to 12 when the SAR =24 and EC= 28 mhos/cm.
8. Estimate the gypsum requirement for given data SAR=22, EC= 320 mmhos/cm, safe valve =10, Depth of irrigation water = 6 cm, purity of gypsum =85%. Equivalent wt. of gypsum= 86.
9. Calculate the gypsum requirement for given data SAR= 25, Ec= 110mmhos/cm , safe SAR= 10, Depth of irrigation water 6cm. Calculate the net quantity of gypsum requirement. Purity of gypsum 90% and equivalent Wt. is 86.
10. Determine the depth of irrigation water, which would change 30 cm depth of loam soil into saline condition, if the EC of irrigation water is 1 mmhos /cm. The density of the soil is 1.2 gm/cm' and the density of water is 1 gm/cm^3. The saturation per centage of the soil is 40.
11. Determine the change in salinity level of the soil due to the evaporation of ground water over a period of four months, when the depth of ground water evaporated is 11 cm, and its electrical conductivity is 10 mmhos/cm. The depth of soil influenced by salt accumulation is 30 cm. The bulk density and saturation per cent of the soil are 1.4 gm/cm^3 and 63, respectively. The density of water is assumed to be 1.0 gm/cm^3.
12. Estimate the leaching requirement when the EC of the saturation extract of the soil is 11 mmhos/cm at 25 per cent reduction in the yield of cotton. The EC of irrigation water is 1.5 mmhos/cm.
13. Estimate the leaching requirement when electrical conductivity (EC) value of the saturation extract of soil is 10 mmhos/cm at 25% reduction in the yield of a crop. The EC of irrigation water is 1.2 mmhos/cm. What will be the required depth of water to be applied to the field if the consumptive use requirement of the crop is 80 mm? EC value of the leaching water may be suitably assumed.

Now, consider that the irrigation requirement for a three-month cropping season is 275 mm and the electrical conductivities of irrigation water and deep percolated waters are 1 and 6 dS/m, respectively. Thus, we get:

$$LR = (1/6)(275) = 46 \text{ mm for three months.}$$

The total water needed to satisfy both irrigation and leaching requirement is 321 mm. Since, irrigation application efficiency is less than 100 per cent, actually more than the above quantity of water is to be applied. If the irrigation application efficiency is 75 per cent, the total amount of water to be applied to satisfy crop water requirement is:

$$(275/0.75) = 367 \text{ mm in three months.}$$

This is more than the sum of the net irrigation requirement and the leaching requirement {(275 + 45) = 320 mm}. The excess water application due to lower than 100 per cent irrigation application efficiency will eventually deep percolate beyond the crop root zone. Hence, the application of a total of 367 mm water in three months will take care of both the irrigation and the leaching requirements.

CHAPTER – 11

SOIL AND WATER QUALITY

11.1 WATER QUALITY

The quality of irrigation water is judged by the amount of suspended and dissolved materials it contains. Suspended materials include soil particles, seeds, leaves and such other debris. The quality of irrigation water gets greatly influenced by its total salt concentration, relative proportion of cations, amount of bicarbonates and of trace elements such as Boron.

The sources of water are as below:

(i) Rivers, or canals taking off from rivers or reservoirs

(ii) Minor drains and lakes

(iii) Ground water through open and bore wells.

- (a) Ground water from arid/semi-arid regions having rainfall less than 500 mm per annum - parts of Rajasthan, Gujarat and parts of Uttar Pradesh are examples.
- (b) Ground water whose quality is influenced by the hydrological condition of high water table caused by river basin irrigation. Certain areas of Punjab, Haryana and Uttar Pradesh are salt affected by rise in water table, and hence therein the ground water is also affected.
- (c) Coastal regions where fresh water aquifers get salt affected due to sea water intrusion, and also low lying ponds and lakes get flooded with sea water during cyclonic storm.

Water is the greatest and universal solvent in nature. It dissolves a formidable range of substances right from the moment a rain drop falls on earth till it joins the sea having flown over or under the ground surface. Sea water could not have been highly salty without this powerful property of water.

Soluble Constituents

Cations which are the most common positively charged ions dissolved in irrigation water are Calcium (Ca), Magnesium (Mg), Sodium (Na) and Potassium (K). The most common negatively charged ions (anions) dissolved in irrigation water are : bicarbonates (HCO_3) sulphates (SO_4) and chlorides (Cl). Other solutes include Carbonate (CO_3), Nitrate (NO_3) and trace elements like Boron (B). Suspended material less than 50 to 100 micron size are often accumulated in channels and pipes and are usually removed physically by dredging.

Salinity

Total concentration of ions in irrigation water is usually called salinity. It is sum of all the ionized (dissolved) salts in water without reference to the specific ions present. The electrical conductivity (EC) of water is often used as a measure of salinity, as the ability of water to conduct electricity is directly related to the number of ions present in it. It is measured as milli-mho per centimetre (m-mho/cm) or decisiemens/metre (dS/ni). One (dS/m) is equal to one (m-mho/cm). dS/m is the SI unit wheareas m-mho/cm is a traditionally used unit. Salinity is also expressed as concentration (mg/l), or part per million (ppm).

The primary effect of salinity is to restrict the availability of soil water to the plant. The presence of salt in soil water increases the energy needed by plants to remove water from soil. Salinity influences crop physiology and its yield. Physiological process like cell enlargement and division, production of proteins and the rate of increase of plant masses are all retarded at high levels of salinity. At extremely high salinity levels, visible injury symptoms like leaf burn are seen.

Toxicity

The most common plant toxins (phytotoxins) found in natural irrigation waters are Sodium (Na), Chloride (CI) and Boron (B). When reclaimed sewage waters are used for irrigation, these are likely to contain many other phytotoxins. Sodium and chloride concentrations in irrigation waters that may not cause toxicity problems in furrow irrigation, may cause toxicity problems in foliar absorption irrigation methods, such as, sprinkler irrigation, that wet the plant leaves, especially when temperature is high and humidity is low. Sprinkler irrigated citrus is known to have been damaged by water even with low concentration of sodium and chloride, though the same level of concentration had no toxic effect when it was irrigated by the same water by furrow or flooding method, in which water does not wet the leaves directly.

11.2 ANALYSIS OF WATER

It is obvious that a chemical analysis of water needs to be carried out to assess its suitability, vis-a-vis, a given crop grown on a given soil. Analyjsis of irrigation water can be broadly carried out by methods that are classified into the three categories as indicated below:

(i) Gravimetric Method

This involves a laborious process. First, the different substances contained in the water samples are to be separated by mechanical means (filtering, decanting, centrifuge method, etc.). Then these are weighed separately. The proportion of each substance with reference to the total sample can then be calculated.

(ii) Volumetric Method

In this method, exact volumes of chemical solutions (of standard strengths) are added to measured parts of the water sample, to bring about specific well established chemical changes - such a volumetric procedure is called titration. The titration point at which the indicative chemical change takes place as the standard solution is being added carefully (in a controlled manner) to the water being analyzed, is the end-point of titration. A few drops of chemical indicators, such a phenolphthalein or methyl-orange serve as indicators of the end point through change of colour. The amount of each of the mineral constituents being analyzed in the water sample is estimated directly from the volume of each of the corresponding standard solution required.

(iii) Colorimetric Analysts

Colorimetric analysis is often used for detecting the presence of and measuring the amount of iron in water. A given volume of water sample is treated with a chemical reagent, and if iron is present, the colour of the sample changes. This colour is then compared with the standard mixtures of water having deferent known concentrations of iron, which are treated with the same reagent resulting in a set of samples with different shades of colour.

It has earlier been mentioned that the sum of all ionized dissolved salts in water determines the salinity level and that the electrical conductivity (EC) stated usually in terms of m-mhos/cm, is a simple measure of salinity. The pH value of natural water (ot soil) is a measure of its alkalinity or acidity. More specifically stated, it is measure of the hydrogen ion-concentration in water. Water molecules (H_2O or HOH) have a mild tendency to break down into ions. Ions which are atoms or grouris of atoms carrying positive (H+) or negative (hydroxyl OH-) charge are formed when water molecules breaks down. In distilled water, the number of H ions formed is such that their concentration is expressed by a pH value of 7. Mathematically, this is the logarithm, to the base 10, of the reciprocal of the hydrogen ion concentration in pure water. Thus, pH = 7 indicates a neutral water, that is neither acidic nor alkaline. A pH of 7.5 to 8.0 usually indicates the presence of carbonates of calcium (Ca) or magnesium (Mg) and a pH of 8.5 and above usually indicates appreciable exchangeable sodium (Na); hence indicating alkalinity.

Water Quality Standards

Soil-plant-water complex has to have an optimum balance between its three constituents. Given a soil matrix for a given crop, water has to have its quality parameters conforming to set standards, or it requires appropriate management practice.

Water Quality Ratings

In the earlier discussion, we have seen that salinity effects are dependent not only on the salt concentration levels in irrigation water, but also on other factors like soil composition, soil texture, water application method and weather conditions like temperature and humidity. Thus, US Soil Salinity Laboratory (USSSL) standards and classification would not be quite suitable to Indian conditions, as in India waters of greater salinity levels than the highest limits proposed by USSSL have been used successfully in some areas of Haryana and Rajasthan. Table 11.1 gives the Water Quality Ratings as per.

Table 11.1. Water Quality Ratings (CSSRI, Karnal, 1972)

Nature of Soil	Crop Type	Safe Max. limit of EC in in-mmhos/cm
Deep black soil and alluvial soil, with clay more than 30% and soil moderately drained	Semi tolerant	1.5
	Tolerant	2.0
Heavy textured soils with clay 20 to 30%; well drained internally, and with good surface drainage system	Semi tolerant	2.0
	Tolerant	4.0
Medium textured soils with clay 10 to 20%, well drained internally and with good surface drainage system.	Semi tolerant	4.0
	Tolerant	6.0
Light textured soils with clay less than 10%; excellent internal and surface drainage	Semi tolerant	6.0
	Tolerant	8.0

11.3. CLASSIFICATION OF WATER

The two most important factors to look for in an irrigation water quality analysis are the Total Dissolved Solids (TDS) and the Sodium Adsorption Ratio (SAR). The TDS of a water sample is a measure of the concentration of soluble salts in a water sample and is commonly referred to as the *salinity* of the water. TDS is expressed in terms of the electrical conductivity (EC) and its units are either:

millimhos per centimeter (mmhos/cm),
deci-Siemens per meter (dS/m) or
micromhos per centimeter (mmhos/cm)

where,

1000 mmhos/cm = 1 mmho/cm = 1 dS/m

The SAR of a water sample is the proportion of sodium relative to calcium and magnesium. Since it is a ratio, the SAR has no units.

Based on Salinity

C1 - Low salinity water -- can be used for irrigation with most crops on most soils with little likelihood that soil salinity will develop. Some leaching is required, but this occurs under normal irrigation practices except in soils of slow and very slow permeability.

C2 - Medium salinity water -- can be used if a moderate amount of leaching occurs. In most cases plants with moderate salt tolerance can be grown without special practices for salinity control.

C3 - High salinity water -- cannot be used on soils with moderately slow to very slow permeability. Even with adequate permeability, special management for salinity control may be required and plants with good salt tolerance should be selected.

C4 - Very high salinity water -- is not suitable for irrigation under ordinary conditions, but may be used occasionally under very special circumstances. The soils must have rapid permeability, drainage must be adequate, irrigation water must be applied in excess to provide considerable leaching, and very salt tolerant crops should be selected.

Based on Sodium

S1 - Low sodium water -- can be used for irrigation on almost all soils with little danger of development of harmful levels of exchangeable sodium.

S2 - Medium sodium water -- will present an appreciable sodium hazard in fine textured soils, especially under low leaching conditions. This water may be used on coarse textured soils with moderately rapid to very rapid permeability.

S3 - High sodium water -- will produce harmful levels of exchangeable sodium in most soils and requires special soil management, good drainage, high leaching, and high organic matter additions.

S4 - Very high sodium water -- is generally unsatisfactory for irrigation purposes except at low and perhaps medium salinity.

11.4. SALT TOLERANT CROPS

Salt tolerance is the ability of the plant to tolerate salt concentration in the root zone. If EC of the saturation extract ECe of the soil is less than 2 m-mho/cm, the effect on the growth and yield is negligible. When ECe is 4 m-mho/cm, the yield of sensitive crops drops down. Most field crops are affected when ECe is between 4 to 8 m-mho/cm. When ECe is between 8 and 16, only high salt tolerant crops would grow. Very few crops grow when the ECe value is above 16 m-mho/cm.

Table 11.2. Relative Salt Tolerance of Crops

	Tolerance		
Crops	**Tolerant**	**Semi-Tolerant**	**Sensitive**
Field Crops	Barley Sugar-beet Tobacco Mustard Cotton Wheat Sugarcane	Rice Sorghum (Jowar) Bajra (Millets) Maize Red gram (Arhar) Green gram (Moong) Sunflower Castor	Beans Bengal gram (Ghana) Pea (Mutter)
Fodder		Lucerne (Alfalfa) Sorghum Maize	
Vegetables	Beetroot Spinach	Tomato Cabbage Cauliflower Potato Carrot	

	Tolerance		
Crops	Tolerant	Semi-Tolerant	Sensitive
		Onion Pea Pumpkin Cucumber Bitter eourd	
Fruits	Coconut Date Palm	Pomegranate Grape Guava Mango Banana Pear Apple Orange Lemon Apricot Peach	

11.5. CLASSIFICATION OF THE SOIL

Salinity and Alkalinity : These terms have earlier bean discussed in the context of properties of irrigation water. They are equally relevant in the context of properties of soils also, since it is through the soil matrix that irrigation water is supplied to the roots of plants, thus influencing the growth and yield of plants. For a better understanding, the following soil-water related terms are defined :

Saline Soil: A soil which has EC of the saturation extract (ECe) more than 4 m-mho/cm at 25°C, ESP less than 15, and pH of the standard paste usually less than 8.5.

Alkaline (or Sadie) Soil: A soil that contains sufficient exchangeable sodium to interfere with the plant growth. ESP of 15 is the lower limit and the pH of the saturated paste is higher than 7, say between 8.5 and 10, and EC less than 4 m-mho/cm.

Saline-Alkaline Soil: A soil with EC of saturation extract greater than 4 m-mhos/cm, and ESP greater than 15 and pH of saturated paste may be 8.5 or slightly more.

Table 11.3. Soil Classification as per pH and EC

pH	Category
6.6 - 7.3	Neutral
7.4 -7.8	Mildly alkaline
7.9 - 8.4	Moderately alkaline
8.5 -9.0	Strongly alkaline
> 9.0	Very strongly alkaline
EC (dS/m)	**Nature of soil**
< 0.8	Normal

0.8 - 1.6	Critical for salt sensitive crops
1.6 - 2.5	Critical for salt tolerant crops
> 2.5	Injurious to all crops

Saturation Extract: This is the solution extracted from the soil at its saturation per centage.

Cation Exchange Capacity (CEC): The total quantity of cations which the soil can absorb by cation exchange is usually expressed as millcquivaleais per 100 gm.

Exchangeable Sodium Per centage (ESP): It is the degree of saturation of the soil which can be effectively exchanged with sodium and is expressed as,

$$ESP = \frac{\text{Exchangeable Sodium (meq per 100 gm of soil)}}{\text{Cation exchange capacity (meq per 100 gm of soil)}}$$

Sodium Absorption Ratio (SAR): It is the ratio used to express the relative activity of sodium ions in exchange reactions with the soil.

Based on the value of EC and SAR, soils are classified into four categories such as: Normal, Saline, Sodic or Alkaline and Saline-alkaline/Saline-sodic.

Table 11.4. Classification of Soil by EC and SAR

Criteria	EC (m-mho/cm)	SAR
Normal	<4	< 13
Saline	>4	< 13
Sodic/Alkaline	<4	> 13
Saline-alkaline	>4	> 13

Note: The SI unit of conductivity is 'Siemens' symbol 'S' per metre. The equivalent non-SI unit is 'mho' and 1 mho = 1 Siemens. Thus for those unused to the SI system mmhos/cm can be read for dS/m without any numerical change.

Conductivity 1 S cm^{-1} (1 mho/cm) = 1000 mS/cm (1000 mmhos/cm)

1 mS/cm^{-1} (1 mmho/cm) = 1 dS/m = 1000 mS/cm (1000 micromhos/cm)

Conductivity to mmol (+) per litre:

mmol (+)/1 = 10 × EC (EC in dS/m)

for irrigation water and soil extracts in the range 0.1-5 dS/m.

Conductivity to osmotic pressure in bars:

OP = 0.36 × EC (EC in dS/m)

for soil extracts in the range of 3-30 dS/m.

Conductivity to mg/l: mg/l = 0.64 × EC x 10^3, or (EC in dS/m) mg/l = 640 × EC for waters and soil extracts having conductivity up to 5 dS/m.

nmol/l (chemical analysis) to mg/l:

Multiply mmol/l for each ion by its molar weight and obtain the sum.

Table 11.5. Soil Salinity Classes and Crop Growth

Soil Salinity Class	Conductivity of the Saturation Extract (dS/m)	Effect on Crop Plants
Non saline	0 - 2	Salinity effects negligible
Slightly saline	2 - 4	Yields of sensitive crops may be restricted
Moderately saline	4 - 8	Yields of many crops are restricted
Strongly saline	8 - 16	Only tolerant crops yield satisfactorily
Very strongly saline	> 16	Only a few very tolerant crops yield satisfactorily

Soil salinity

The salt concentration in the water extracted from a saturated soil (called saturation extract) defines the salinity of this soil. If this water contains less than 3 grams of salt per litre, the soil is said to be non saline (see Table below). If the salt concentration of the saturation extract contains more than 12 g/l, the soil is said to be highly saline.

Table 11.6. Salt concentration and salinity classes

Salt concentration of the soil water (saturation extract)		Salinity
in g/l	in millimhos/cm	
0 - 3	0 - 4.5	non saline
3 - 6	4.5 - 9	slightly saline
6 - 12	9 - 18	medium saline
more than 12	more than 18	highly saline

11.6 DETERMINATION OF CHEMICAL PROPERTIES OF SOIL AND WATER

Preparation of Saturation Extract

Saturated soil paste is prepared by adding distilled water to a sample of soil while stirring with a spatula. After mixing, the sample should be allowed to stand for an hour or more, and then the criteria for saturation should be rechecked. Free water should not collect on the soil surface nor should the paste stiffen markedly or lose its glistening appearance on standing.

1. Determination of pH

pH meter consists of two electrodes:

(i) Glass electrode or indicator electrode and

(ii) Calomel electrode or reference electrode.

When the two electrodes are dipped into aqueous solution under test, the potential is develops in the solution. The difference in potential between both the electrodes is measured by pH meter.

Apparatus

- Glass electrode
- pH meter
- 50 ml beaker
- volumetric cylinder and glass rod

Preparation of Buffer Solution (Ph 7.0, 9.2, 4.0)

The buffer tablets of standard pH are available. Particular tablet is to be dissolved in freshly double distilled water and made upto 100 ml. It should be stored in polythelene or pyrex glass container.

Procedure

- Take 10 g soil in 50 ml beakers. Add 20 ml of distilled water and stir intermittently with glass rod for 30 minutes.
- Determine pH of the soil suspension with the pH meter as per direction with the instrument. Stir the suspension in each beaker with a glass rod again just before taking the pH reading.
- Wash the electrodes with distilled water after each determination.
- Express pH to the nearest tenth of pH unit.

2. Determination of Electrical Conductivity of Soil

The choice of a method to measure groundwater salinity depends on the reason for making the measurements, the size of the area and the time and the budget available for doing the work.

Once the network of observation wells and boreholes has been set out, a representative number of water samples is taken. Sampling can often best be combined with other drainage investigations, such as measuring hydraulic conductivity in open boreholes.

The salinity of groundwater can be rapidly determined by measuring its electrical conductivity (EC).

Electrical conductivity is expressed in deciSiemens per metre (dS/m), formerly in millimhos per centimetre (mmhos/cm). Expressing the results in terms of specific electrical conductivity makes the determination independent of the size of the water sample. Conductivity cannot simply be related to the total dissolved solids because groundwater contains a variety of ionic and undissociated species. An approximate relationship for most groundwater with an EC-value in the range of 0.1 to 5 dS/m is: 1 dS/m 640 mg/1.

The EC expresses the total concentration of soluble salts in the groundwater, but gives no information on the types of salts. These may be calcium, magnesium, sodium, potassium, carbonate, bicarbonate, chloride, sulphate, and nitrate, and need to be determined in the laboratory. Since these chemical analyses are costly, not all the observation points need be sampled for detailed analysis. A selection of sites should be made, based on the results of the EC-measurements.

The solution offers some resistance to the passage of electrical current through them depending upon the concentration and the type of ions present. Higher the salt content, lesser the resistance to the flow of current. The resistance (R) is defined by Ohm's law as the ratio of electrical potential in volts (E) and strength of current in ampere (I).

$$E/I = \text{Volts/ Current} = R \text{ in Ohm}$$

Electrical conductivity or conductance is the reverse of resistance.

$$C = 1/R$$

$$= 1/\text{Ohm} = \text{mho (reverse of Ohm)}$$

Apparatus

- Conductivity meter and a conductivity cell with known cell constant,
- Beaker 50 ml,
- Volumetric cylinder,
- Glass rod, etc.

Reagents

(1) Saturated solution of calcium sulphate (reagent quality)

(2) 0.01 N KCl solution: Dissolve 0.7456 g of potassium chloride in distilled water and dilute to 1 liter.

Procedure

- Calibrate instrument at 0 and 100 or as per directions supplied with the instrument.
- Check the instrument with the saturated calcium sulphate (EC = 2.2 mhos/cm) or 0.01 N KCI solution (EC = 1.41 dSm^{-1}) before proceeding with the samples.
- The same soil suspension prepared for determination of pH is also used for salinity determination, after recording the soil pH.
- Allow the soil suspension in the beaker to settle (or take 10 g soil in 50 ml beaker and add 20 ml of distilled water and stir with glass rod intermittently for 30 minutes).
- Pour the supernatant liquid into the conductivity tube after rinsing it with the same supernatant solution.
- Dip the conductivity cell and read on direct conductivity meter.
- Express conductivity in dSm^{-1} (mmhos/cm) at 25°C to the nearest first decimal.

11.7. CONJUNCTIVE USE

Blending

Where drainage water salinity exceeds the threshold values for optimal crop production, it can be mixed with other water resources to create a mixture of acceptable quality for the prevailing cropping patterns.

Where reuse takes place by mixing drainage water from main drains with surface water in main irrigation canals, the most salt sensitive crop determines the final water quality. For example, according to the regulations for the Nile Delta, the maximum salinity of the blended water is rather low to ensure optimal production of the major crops grown, i.e. cotton, maize, wheat, rice and berseem. Maize and berseem are the most salt sensitive crops with an EC threshold of 1.7 and 1.5 dS/m, respectively (Maas and Grattan, 1999). To ensure potential maximum yield production of these two crops and assuming a concentration factor of two between the EC of the irrigation water and the EC, the maximum allowable salinity of the irrigation water is about 0.8-0.9 dS/m, Where mixing takes place at the farm level, the salinity of the blended water can be adjusted towards the salt tolerance of individual crops. All irrigations applied with a blended water mixture having a maximum salinity of 3 dS/m, a potential crop yield of 90 per cent of the maximum yield was obtained. According to the crop tolerance data published by Maas and Grattan (1999) and assuming a concentration factor of 1.5 between water salinity and soil salinity, the maximum potential yield could be attained. The difference between theory and the actual field situation might be caused by a discrepancy in the actual concentration factor, which is the inverse of the LF. Moreover, the actual yields depend on many factors including physical, climate and farm management practices. In general, where all irrigations use the blended water, the maximum allowable salinity depends on the threshold value of the crops. Where the pre-irrigation uses freshwater, the blended water could have a salinity level exceeding the threshold value without affecting yields. The experiment from India shows that water used in post-plant irrigations with a salinity level of three times (ECI = 9 dS/m) the maximum allowable salinity level when only blended water is used (ECI = 3 dS/m) results in 90 per cent yields.

Table 11.7. Drainage Water Quality Criteria for Irrigation Purposes

Salinity of drainage water (dS/m)	Restriction on use for irrigation
< 1.0	used directly for irrigation
1.0 - 2.3	mixed with canal water at ratio 1:1
2.3 - 4.6	mixed with canal water at ratio 1:2 or 1:3
> 4.6	not used for irrigation

Table 11.8. Effect of diluted drainage water on Wheat Yield

ECI, dS/m	Relative Yield (%)	
	All irrigations	Post-plant irrigations
0.5 (canal water)	100	100
3.0	90.2	-
6.0	80.4	95.8
9.0	72.5	90.3
12.0	56.4	83.7
18.0	-	78.0

Source: Sharma *et al.*, 2000.

The leaching of salts is necessary to maintain a favourable salinity balance in the rootzone. The procedures presented above might be used for this purpose. On the other hand, if higher salinity levels are tolerated towards the end of the growing period or if a more salt sensitive crop is grown after a more salt tolerant crop, the salinity levels have to be reduced sufficiently low in order not to interfere with the growth of the next crop. Equation 19 could be used to calculate the required amount of leaching.

11.8. CYCLIC USE

Cyclic use, also known as sequential application or rotational mode, is a technique that facilitates the conjunctive use of freshwater and saline drainage effluent. In this mode, saline drainage water replaces canal water in a predetermined sequence or cycle. Cyclic use is an option for where the salinity of the drainage water exceeds the salinity threshold value of the desired crop. A condition for cyclic use is that two different water sources can be applied to the field separately.

Table 11.9. Plants and Irrigation water salinity

Plants	**Irrigation water salinity (ECw) (fresh, blended, or drainage)**
Salt-sensitive vegetables	below 2 dS/m
Salt-tolerant vegetables and flowers	below 4.5 dS/m
Salt-tolerant field crops *(cotton, canola, barley, sugarbeet)*	below 7 dS/m
Salt-tolerant forages	8-14 dS/m
Halophytes	15 dS/m and above
Salt-tolerant trees	5-10 dS/m

CHAPTER – 12

WATER AND SALT BALANCE OF AGRICULTURAL LAND

12.1 WATER BALANCE IN AGRICULTURE LAND

The water balance is based on the principle of the conservation of mass for boundaries defined in space and time and can be written as:

$$\text{Inflow} = \text{Outflow} + \Delta W$$

Where

ΔW is the change in water storage

When the change in storage is positive, the water content increases and, when negative (i.e. there is depletion instead of storage), it decreases.

The ΔW values of the different reservoirs can be indicated by ΔW_s, ΔW_r, ΔW_x and ΔW_q respectively.

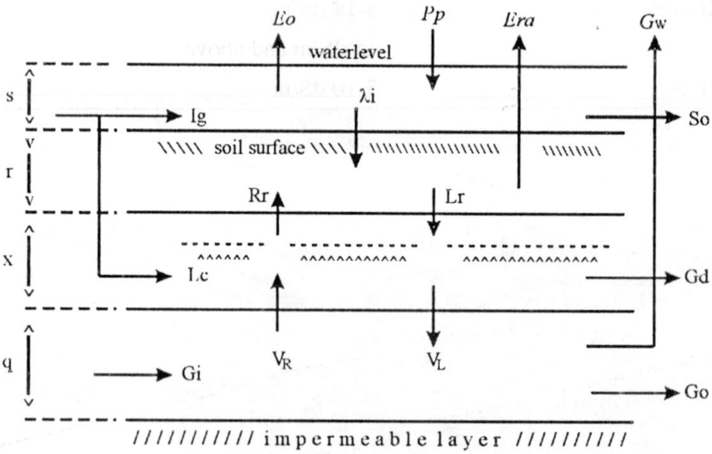

Figure 12.1. The concept of 4 reservoirs with hydrological inflow and outflow components

In Figure the soil profile is divided into four different reservoirs.

s: Reservoir above the soil surface (surface reservoir)

r: Reservoir below the soil surface from which evapotranspiration takes place. Normally, this reservoir is taken equal to the root zone.

x: Reservoir of the transition zone between root zone and aquifer. Its lower limit can be fixed in different ways according to local conditions:

1 - at the interface between a clay layer on top of a sandy layer, provided that the water table remains above the interface

2 - at the annually deepest depth of the water table

3 - at the deepest depth to which the influence of a subsurface drainage system extends

4 - at the depth where horizontal groundwater flow is converted into vertical flow of groundwater or vice versa.

q: Reservoir of the aquifer resting on an impermeable layer.

Each reservoir has incoming and outgoing groundwater factors as shown in the figure.

List of Symbols for Water Balance

E_o : Evaporation from open water on the soil surface.

E_{ra} : Actual evapo-transpiration from the root zone. When the root zone is moist, it equals the potential evapo- transpiration E_p, otherwise it will be less.

G_d: Subsurface drainage, natural (Gdn) or artificial (Gda), to channels, ditches or pipe systems. The suffixes "dn" and "da" are not used in the given water balances, but can be introduced.

G_i: Horizontally incoming groundwater into the aquifer.

G_o: Horizontally outgoing groundwater from the aquifer.

G_w: Pumping from wells placed in the aquifer.

I_g: Horizontally incoming surface water. This can consist of natural inundation and/or gross surface irrigation.

L_c: Infiltration, from river or canal systems into the transition zone, often referred to as seepage losses from canals.

L_r: Percolation of water from the unsaturated root zone into the transition zone.

λ_i: Infiltration of water through the soil surface into the root zone.

P_p: Downwardly incoming water to the surface: precipitation (incl. snow), rainfall, sprinkler irrigation.

R_r: Capillary rise of water from the transition zone into the unsaturated root zone, i.e. the water table is below the root zone.

S_o: Surface runoff (natural, S_{on}) or surface drainage (artificial, S_{oa}). The suffices "on" and "oa" are not used in the given water balances, but can be introduced.

V_R: Vertically upward seepage of water from the saturated aquifer into the transition zone.

V_L: Vertically downward drainage of water from the saturated transition zone into the aquifer.

12.2 WATER BALANCE OF THE SURFACE RESERVOIR

The surface storage or reservoir(s) is located on top of the soil. The water balance of the surface reservoir for a certain period reads:

$$P_p + I_g = E_o + \lambda_i + S_o + \Delta W_s$$

Where,

P_p: the amount of water vertically reaching the soil surface, such as precipitation and sprinkler irrigation,

Ig : the gross irrigation inflow including the natural surface inflow and the drain and well water used for irrigation, but excluding the percolation losses from the canal system,

Eo : the amount of evaporation from open water,

λi : the amount of water infiltrated through the soil surface into the root zone, So is the amount of surface runoff or surface drainage leaving the area, and

ΔWs: the change in amount of water stored in the surface reservoir.

12.3. WATER BALANCE IN THE ROOT ZONE

The root zone (r) corresponds to the depth of soil from which evapotranspiration takes place. Its water balance reads:

$$\lambda_i + R_r = E_{ra} + L_r + \Delta W_r$$

Where,

R_r : the amount of capillary rise into the root zone,

Era : the amount of actual evapo-transpiration from the root zone,

Lr : the amount of percolation losses from the root zone,

ΔWr: the storage of water in the root zone.

The factor R_r is the opposite of L_r and these components cannot occur simultaneously, i.e. when $R_r > 0$ then $L_r = 0$ and vice versa.

When water balances are made for fairly long periods of time, for instance a season or a year, the storage ΔW_r is often negligibly small compared to the other hydrological components. Therefore, this storage is set equal to zero and the water balance changes to:

$$\lambda_i + R_r = E_{ra} + L_r$$

In the root zone, there are only vertical, no horizontal water movements.

12.4. WATER BALANCE IN THE TRANSITION ZONE

The transition zone (x) is the zone between root zone and aquifer. Its lower limit can be fixed in different ways according to local conditions: (a) at the interface between a clay layer on top of a sandy layer, (b) at the annually greatest depth to water table, (c) at the greatest depth to which the influence of a subsurface drainage system extends, (d) at the depth where horizontal groundwater flow is converted into vertical flow of groundwater or vice versa. The water balance of the transition zone, reads:

$$L_r + L_c + V_r = R_r + V_L + G_d + \Delta W_x$$

Where

L_c: the percolation loss from the irrigation canal system,

VR: the amount of vertical upward seepage from the aquifer into the transition zone,

VL : the amount of vertical downward drainage from the saturated transition zone to the aquifer,

Gd : the total amount of natural or artificial drainage of groundwater to ditches or pipe drains, and

ΔWx : the water storage in the transition zone.

The component V_R is the opposite of V_L and these cannot occur simultaneously, i.e. when $V_R > 0$ then $V_L = 0$ and vice versa.

12.5. WATER BALANCE IN THE AQUIFER

The water balance of the aquifer (q) can be written as:

$$G_i + V_L = G_o + V_R + G_w + \Delta W_q$$

where:

G_i : the amount horizontal groundwater inflow through the aquifer,

Go : the amount of horizontal groundwater outflow through the aquifer,

Gw : the amount groundwater pumped from the aquifer through wells,

ΔWq : the groundwater storage in the aquifer.

When the aquifer is not elastic (compressible or de-compressible) and the aquifer is saturated, the value of ΔW_q is zero.

12.6. COMBINED WATER BALANCES

When the water table is in the transition zone, the balances of the surface reservoir and the root zone may be combined in to the topsoil water-balance, by adding balances

$$E_a = E_o + E_{ra}$$

where

E_a : the total actual evapo-transpiration.

This gives:

$$P_p + I_g + R_r = E_a + S_o + L_r + \Delta W_s + \Delta W_r$$

In the topsoil water-balance, the infiltration component λ_i is not present. This component is a vertical flow linking the two reservoirs. It is called a linkage component. Using:

$$I_f = I_g - S_o$$
$$V_s = P_p + I_f$$

where

I_f : the net field irrigation and

V_s represents the net surface water resource:

$$V_s + R_r = E_a + L_r + \Delta W_r + \Delta W_s$$

With a water table in the transition zone, the balances of the transition zone and aquifer can be combined into the geo-hydrologic water balance, in which the storage ΔW_q may be considered zero as the aquifer is fully saturated:

$$L_r + L_c + G_i = R_r + G_o + G_d + G_w + \Delta W_x$$

Here, the linkage components V_R and V_L have vanished.

It is also possible to combine three reservoirs. For example, joining the water balances of the surface reservoir, root zone and transition zone, gives the agronomic water-balance:

$$P_p + I_g + L_c + V_R = S_o + E_a + G_d + V_L + \Delta W_s + \Delta W_r + \Delta W_x$$

Here, the linkage components R_r and L_r have vanished.

When the water table is not in the transition zone, it may be above the soil surface, in the root zone, or in the aquifer. The balances can be adjusted accordingly as discussed below.

Water Table above the Soil Surface

When the water table remains above the soil surface, the values of ΔW_r, ΔW_x and ΔW_q are zero, as the soil is fully saturated. When, in addition, the water flows from the subsoil into the surface reservoir, the infiltration λ_i becomes negative. Thus, it is preferable to combine the water balances of all the reservoirs:

$$P_p + I_g + L_c + G_i = E_a + S_o + G_o + G_d + G_w + \Delta W_s$$

In this overall water balance, all linkage components have disappeared.

Water Table in the Root Zone

When the water table is in the root zone, the capillary rise R_r and percolation L_r do not exist, because the transition zone is saturated. Also, the values of ΔW_x and ΔW_q are zero. Thus it is preferable to combine the water balances of root zone, transition zone and aquifer, giving the subsoil water balance:

$$\lambda_i + L_c + G_i = E_{ra} + G_o + G_d + G_w + \Delta W_r$$

Water Table in the Aquifer

When the water table is in the aquifer, the saturated, vertical, flows into or from the transition zone (V_R and V_L) do not exist. In this case one can use either the geo-hydrologic or the sub soil water balances.

Reduced number of reservoirs

When one of the four reservoirs is not present, the balances become simpler. For example, when there is no aquifer, the geo- hydrologic water-balance no longer exists as it will be reduced to the transition zone balance with $V_L = 0$ and $V_R = 0$. Also, the subsoil water-balance will be reduced to:

$$\lambda_i + L_c = E_{ra} + G_d + G_w + \Delta W_r$$

and the overall water balance to:

$$P_p + I_g + L_c = E_a + S_o + G_d + G_w + \Delta W_s$$

When two or more reservoirs are absent, the water balance equations become still more simple.

12.7 STEADY STATE CONDITION

When the water balances are made for fairly long periods (e.g. per season or per year), in many cases the ΔW values (changes in storage) are small compared to the values of the other factors (components) of the water balance. Then, the ΔW values can be ignored, so that the water balances can be simplified. When Δ is taken zero, i.e. the amounts of all incoming and all outgoing water are equal, one obtains a steady state.

Therefore, over fairly long periods of time, the water balance can often be considered in steady state. For example, the water balance of the transition zone in steady state is:

$$L_r + L_c + V_r = R_r + V_L + G_d$$

The geo-hydrologic water balance in steady state is:

$$L_r + L_c + G_i = R_r + G_o + G_d + G_w$$

while the agronomic water balance in steady state is:

$$P_p + I_g + L_c + V_R = S_o + E_a + G_d + V_L$$

The overall water balance in steady state will be

$$P_p + I_g + L_c + G_i = E_a + S_o + G_o + G_d + G_w$$

and the subsoil water balance in steady state becomes:

$$\lambda_i + L_c + G_i = E_{ra} + G_o + G_d + G_w$$

All the above examples of steady state balances contain the subsurface drainage component (Gd), so there are different expressions available to calculate this component depending on the availability of data.

Net and excess values

Linkage factors along the same reservoir boundary having arrows pointing in opposite direction, especially percolation L_r versus capillary rise R_r and upward seepage V_R versus downward drainage V_L, cannot occur at the same time but they can occur alternately in different periods of time. Over a longer period of time, one factor will be greater than the other and they can be combined into net values. For example:

$$R_{rn} = R_r - L_r \qquad \text{Net capillary rise}$$

$L_{rn} = L_r - R_r$ Net percolation

Note that $R_rN>0$ when $L_rN=0$ and $L_rN>0$ when $R_rN=0$.

$V_{Rn} = V_R - V_L$ Net upward seepage

$V_{Ln} = V_L - V_R$ Net downward drainage

Note that $V_{Rn}>0$ when $V_{Ln}=0$ and $V_{Ln}>0$ when $V_{Rn}=0$.

The horizontal inflow (G_i) and outflow (G_o) of groundwater, which have arrows pointing in the same direction can be combined into excess values:

$G_iN = G_i - G_o \quad > 0$

Excess inflow over outflow

$G_oN = G_o - G_i \quad > 0$

Excess outflow over inflow Note that $G_iN>0$ when $G_oN = 0$ and $G_oN>0$ when $G_iN=0$.

With the net and excess values, the previous balances can be simplified further. For example, the steady state balance of the aquifer (with $\Delta W_q = 0$), can now be written as:

$G_iN = V_{Rn} + G_w$ or: $G_iN = G_w - V_{Ln}$ or: $G_oN = V_{Ln} - G_w$

12.8 EXAMPLES OF WATER BALANCES

Example

An example of the surface water balance is given in Figure, giving the relation between surface runoff and rainfall. The principle illustrated in the figure is used in the Curve Number Method.

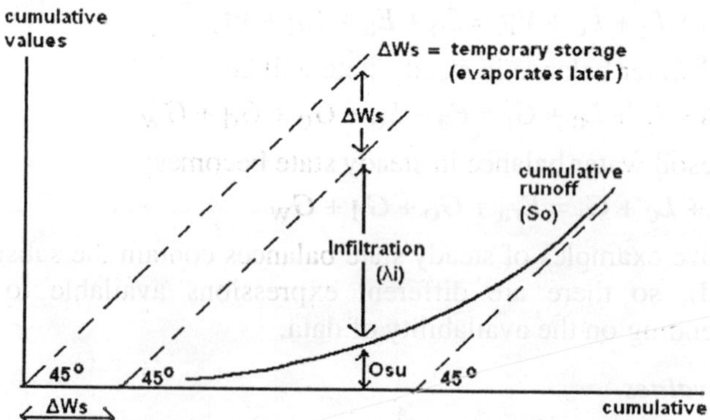

Figure 12.2. Illustration of a surface water balance during periods of high rainfall

Figure Illustration of a surface water balance during periods of high rainfall: So = Pp – λi - ΔWs with ΔWs = Eo after rain has ceased

Example

Another example, based on climatic data, is presented in Figure. It concerns a simplified topsoil water balance in which I_g, E_o, S_o, R_r and ΔW_s are taken equal to zero:

$$P_p = E_{ra} + L_r + \Delta W_r$$

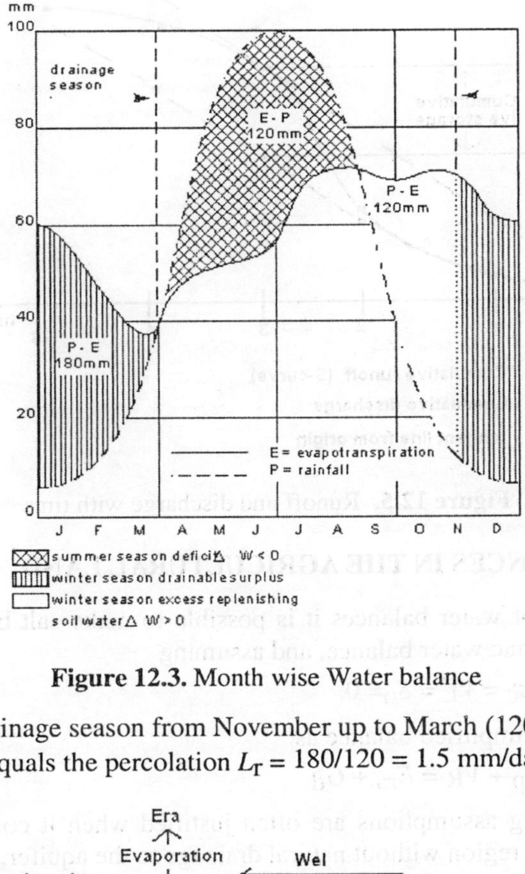

Figure 12.3. Month wise Water balance

Fixing the drainage season from November up to March (120 days), the average drainable surplus equals the percolation L_r = 180/120 = 1.5 mm/day.

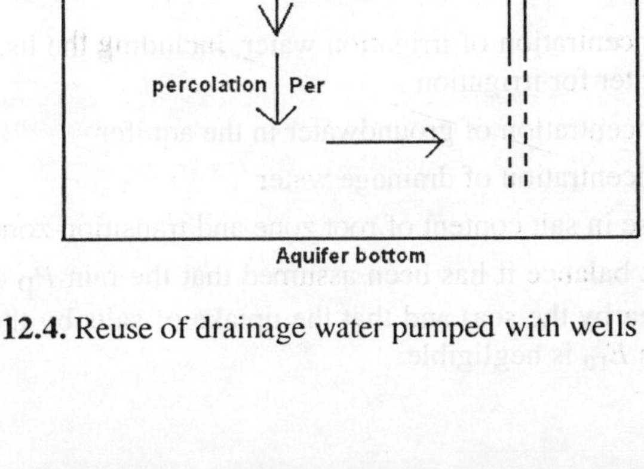

Figure 12.4. Reuse of drainage water pumped with wells for irrigation

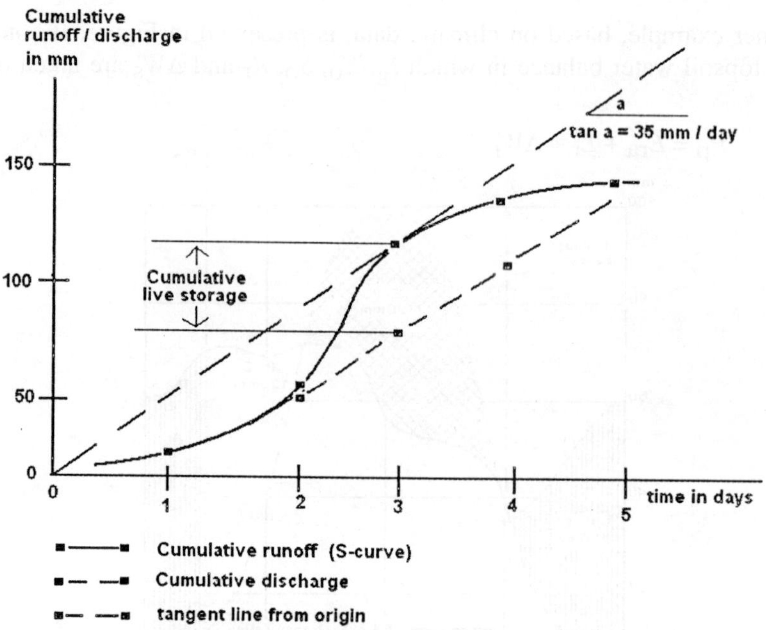

Figure 12.5. Runoff and discharge with time

12.9. SALT BALANCES IN THE AGRICULTURAL LAND

On the basis of water balances it is possible to make salt balances. Using the steady-state agronomic water balance, and assuming

$$E_o = L_c = V_L = S_o = 0,$$

we obtain the simplified balance as:

$$I_g + P_p + V_R = E_{ra} + G_d$$

The simplifying assumptions are often justified when it concerns an irrigated area in a (semi) arid region without natural drainage to the aquifer.

The salt balance is obtained by multiplying the water balance factors with their respective salt concentrations:

$$IgCir + VRCq = GdCd + \Delta Zrt$$

Where,

Cir: salt concentration of irrigation water, including the use of drain and/or well water for irrigation

Cq: salt concentration of groundwater in the aquifer

Cd: salt concentration of drainage water

ΔZrt : increase in salt content of root zone and transition zone

In the above salt balance it has been assumed that the rain P_p contains no salts (i.e. the area is not nearby the sea) and that the uptake of salts by the plants through the evapotranspiration E_{ra} is negligible.

Stating that $\Delta Zrt = 0$ (i.e. allowing no increase in salt content) one finds:

$$Ig = GdCd - VRCq / Cir$$

Using the same simplifying assumptions, the steady-state water balance of the transition zone can be written as:

$$G_d = L_r - R_r + V_R$$

The corresponding salt balance is:

$$G_dC_d = L_{rn}C_L + V_RC_q$$

where

$L_{rn} = L_r - R_r$ (net percolation, balance) and

C_L = the salt concentration of the percolating water.

Combining salt balances yields:

$$I_g = L_{rn}C_L/C_{ir}$$

The salt concentration (C_L) of the percolation or leaching water is a function of the salt concentration of the soil moisture in the root zone:

$$C_L = f(C_r)$$

This is the leaching efficiency function. The function represents the effects of the heterogeneity of the soil in vertical and horizontal direction, and the subsequent irregular passage of water and salt through the soil.

Therefore,

$$I_g = L_{rn}f(C_r)/C_{ir}$$

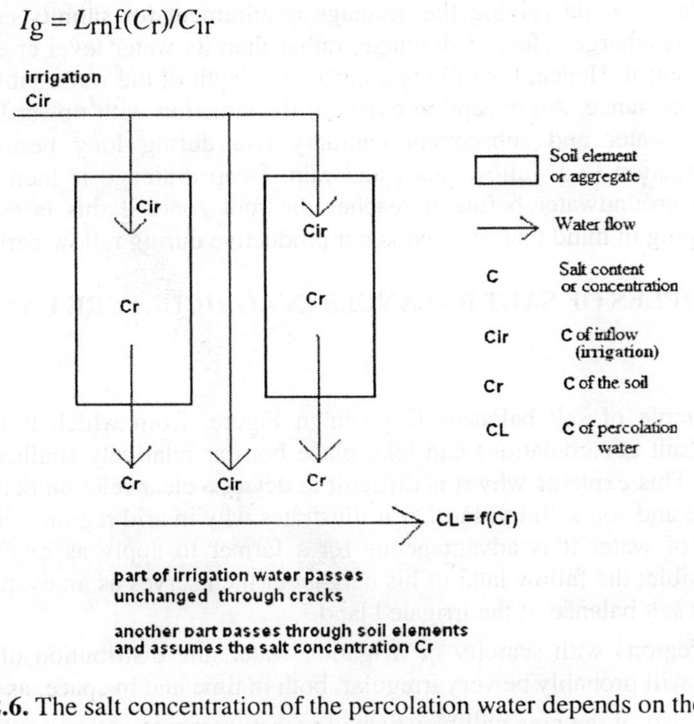

Figure 12.6. The salt concentration of the percolation water depends on the degree of mixing of infiltration water with the soil aggregates.

Further, using the simplifying assumptions mentioned before, the topsoil water balance in steady state yields:

$$L_{rn} = I_g + P_p - E_{ra}$$

The last two equations combined give:

$$I_g = (I_g + P_p - E_{ra}).f(C_r) / Cir$$

After rearrangement, to make the previous equation explicit in I_g, it can be changed into:

$$I_g = (E_{ra} - P_p)\ f(C_r) / f(C_r) - C_{ir}$$

This is the irrigation requirement for salinity control.

From the above formula it is seen that, to maintain a certain permissible salt concentration of the root zone

($C_r = C_{rp}$), the irrigation I_g should be more than $E_{ra} - P_p$ (i.e. there should be excess irrigation) and that more excess irrigation is required as its salt concentration (C_{ir}) increases. Drainage for salinity control, therefore, can only be helpful if the irrigation is appropriate. The corresponding drainage discharge can be determined from the agronomic water balance in steady state as:

$$G_d = I_g + P_p + V_R - E_{ra}$$

or, after substitution of Equation:

$$Gd = [(Era - Pp)\ f(Cr) / f(Cr) - Cir\ ] + P_p + V_R - E_{ra}$$

From this formula (giving the drainage requirement for salinity control) it is seen that the discharge effect of drainage, rather than its water level effect, governs the salinity control. Hence, for salinity control, the depth of the water table is only of secondary importance. An exception exists in the situation with upward seepage of salty ground water and subsequent capillary rise during long periods without sufficient irrigation (e.g. fallow land) and rain. Deep drainage is then required to intercept the groundwater before it reaches the root zone, if this is economically feasible, keeping in mind that the land is not productive during fallow periods.

12.10 EXAMPLES OF SALT BALANCES IN AGRICULTURE LAND

Example

An example of salt balances is given in Figure, from which it is seen that salinization (salt accumulation) can take place both at relatively shallow and deep water tables. This explains why it is difficult to detect a clear relation between depth of water table and soil salinity. Further it illustrates why in arid regions with flat land and scarcity of water it is advantageous for a farmer to apply as much irrigation water as possible; the fallow land of his neighbour then serves as an evaporation pan favouring the salt balance of the irrigated land.

In dry regions with scarcity of irrigation water, the distribution of this water over the area will probably be very irregular, both in time and in space, as follows:

(a) A part of the area will be irrigated with excess water; there will be sufficient leaching; salinity problems are absent; the deep percolation losses will disappear through seepage, capillary rise and evaporation in neighbouring fallow land;

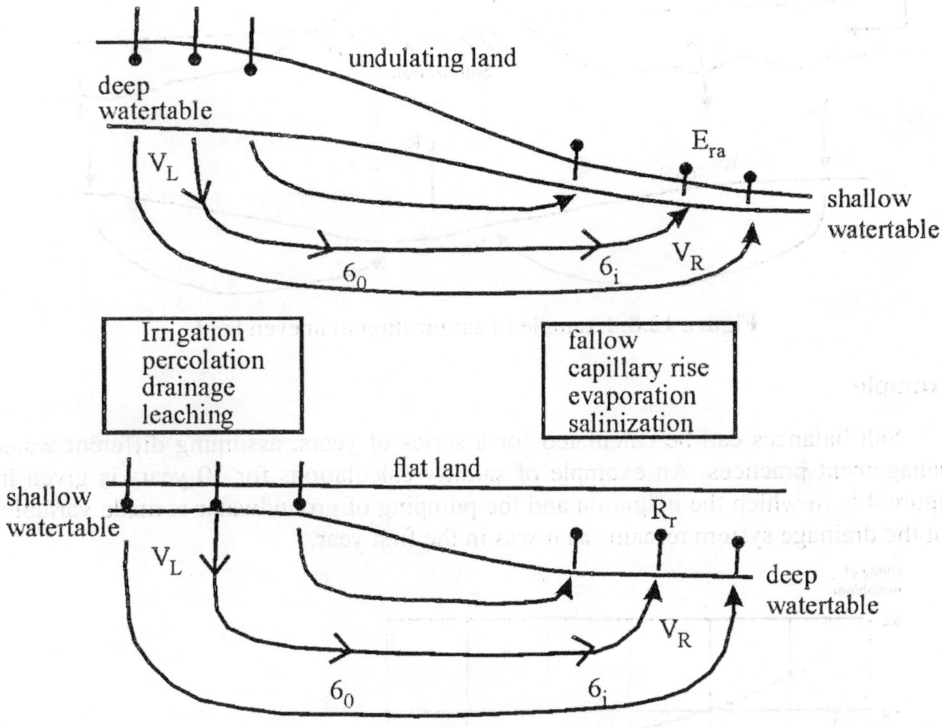

Figure 12.7. Examples of the relation between hydrology and salinization.

(b) A part of the area will be irrigated with an insufficient amount of water; no leaching will take place and salinity problems will develop, even when there is no upward seepage of groundwater.

(c) A part of the area is not irrigated (fallow); salinity problems may or may not develop dependent on the presence or absence of upward seepage;

(d) Some parts of the area are permanently irrigated, other parts are only seasonally irrigated, and the remaining parts are either irrigated once in a number of years or hardly ever.

Efficient salinity control in areas with scarce irrigation water, therefore, is primarily a matter of efficient management of irrigation water. Drainage is only a complementary activity. It would make little sense to drain fields that are not irrigated yearly or that have no drainable surplus even when irrigated. Sometimes it is advisable not to install a drainage system at all, but to use the permanently fallow land as "evaporation pan" for the drainage water from irrigated fields, if this is socially acceptable. In the irrigation management, not only the application of the correct amount of irrigation is important but also the land levelling.

This is illustrated in Figure showing poorly levelled land.

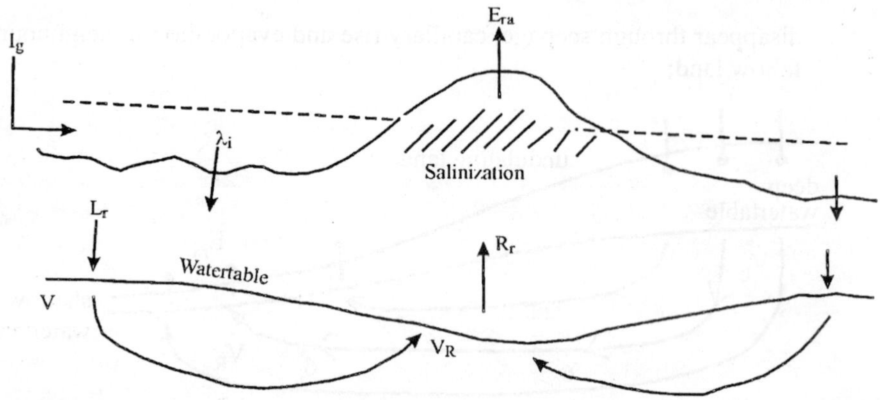

Figure 12.8. Example of salinization in uneven land

Example

Salt balances can be calculated for a series of years, assuming different water management practices. An example of salinity calculations for 30 years is given in Figure 4.5, in which the irrigation and the pumping of groundwater is made variable, but the drainage system remains as it was in the first year.

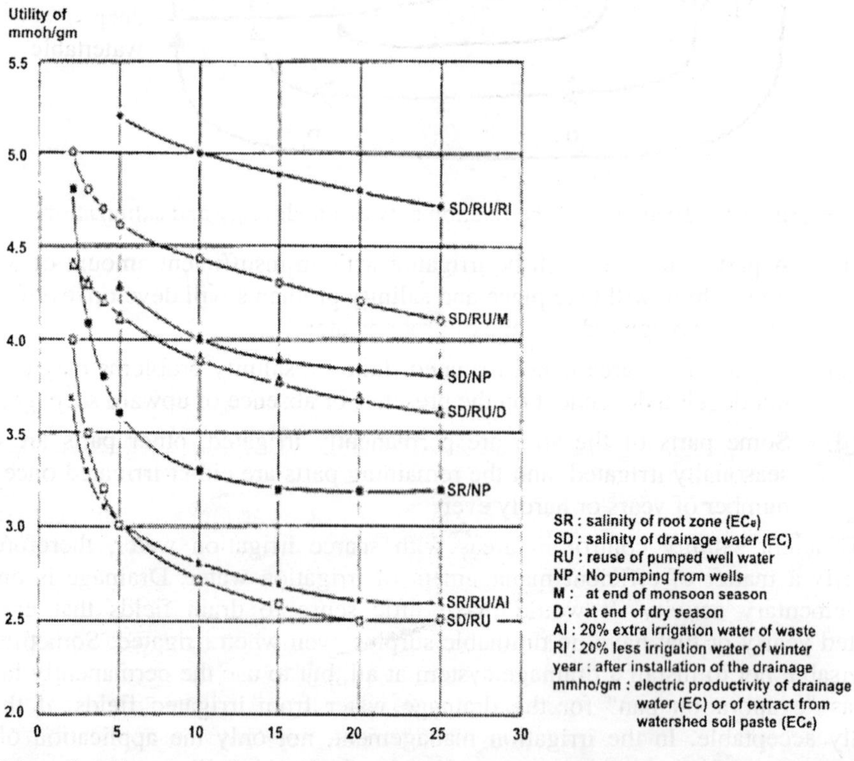

Note that the symbols used in the figure deviate from those used previously.

The figure shows that 20% more irrigation water during the dry winter season would accelerate the desalinization of the root zone (line SR/RU/AI). Conversely

20% less irrigation water reduces the speed of desalinization considerably (upper line SR/RU/RI). The salinity of the root zone is higher at the end of the hot monsoon season than at the end of the winter season (compare lines SR/RU/M and SR/RU/D). Apparently the amount of rainfall during the monsoon season is not sufficient to prevent a slight resalinization. Further the figure shows that the use of pumped groundwater retards the desalinization (compare lines SR/RU/M and SR/NP), but it speeds up the reduction of the salt concentration in the drainage water (compare lines SD/RU and SD/NP).

Example

In an area there is an irrigation season of 100 days followed by a rain fed cropping season (265 days). The water balance factors during these periods are shown in Figure and in the following table.

Table 12.1. Water balance factors

Irrigation season				**Rain fed season (un-irrigated)**			
P_p	=	100	mm	P_p	=	500	mm
E_{ra}	=	500	mm	E_{ra}	=	800	mm
V_R	=	100	mm	V_R	=	200	mm
I_g	=	?	mm	I_g	=	0	mm
ΔW_r	=	?	mm	ΔW_r	=	?	mm
L_r	=	?	mm	L_r	=	?	mm
R_r	=	?	mm	R_r	=	100	mm
G_d	=	?	mm	G_d	=	?	mm
C_{ir}	=			EC	=	1	mmho/cm or dS/m

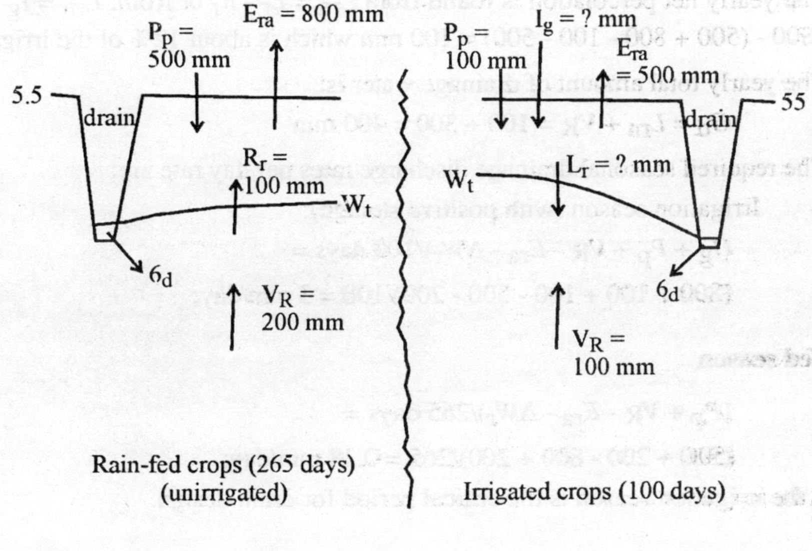

Rain-fed crops (265 days) (unirrigated) | Irrigated crops (100 days)

The leaching efficiency function $f(C_r)$ is taken simply equal to C_{rp} (i.e. the salt concentration of the percolation water equals the permissible salt concentration C_{rp} of the soil moisture in the root zone), and the permissible salt concentration of the soil moisture in the root zone is fixed as

$$EC = C_r = C_{rp} = 8 \text{ mmho/cm or dS/m.}$$

Applying equation using the data of the whole year we obtain the irrigation requirement for salinity control as:

$$I_g = (500 + 800 - 100 - 500) \times 8/(8 - 1) = 800 \text{ mm}$$

This amount of irrigation water is required to:

- cover the consumptive use of the crop in the irrigation season:

  $E_{ra} - P_p = 500 - 100 = 400$ mm;

- replenish the soil moisture that has been used by the crops in the non-irrigation season:

  $\Delta W_r = E_{ra} - P_p - R_r = 800 - 500 - 100 = 200$ mm;

- leach the salts that have been brought in by the capillary rise in the non-irrigation season;
- leach the salts that have been brought in with the irrigation water.

Since 1 mmho/cm or 1 dS/m (the salt concentration of the irrigation water) corresponds to roughly 0.6 g salt/l water =

0.6 kg salt per m³ water, and since per ha annually

0.800 m x10000 m² = 8000 m³ of irrigation water is applied per ha, this means that actually

0.6 kg/m³ x 8000 m³ = 4800 kg of salt per ha is introduced to the land by irrigation, and these salts need also to be leached annually if no salt accumulation is to take place ($\Delta Z_{rt} = 0$).

The yearly net percolation is found from $L_{rn} = L_r - R_r$ or from: $L_{rn} = I_g - (E_{ra} - P_p) = 800 - (500 + 800 - 100 - 500) = 100$ mm which is about 12% of the irrigation.

The yearly total amount of drainage water is:

$$G_d = L_{rn} + V_R = 100 + 300 = 400 \text{ mm}$$

The required seasonal drainage discharge rates per day rate are:

– Irrigation season (with positive storage):

  $(I_g + P_p + V_R - E_{ra} - \Delta W_r)/100$ days =

  $(800 + 100 + 100 - 500 - 200)/100 = 3$ mm/day;

Rain fed season

$(P_p + V_R - E_{ra} - \Delta W_r)/265$ days =

$(500 + 200 - 800 + 200)/265 = 0.38$ mm/day;

so that the irrigation season is the critical period for drain design.

Note : The permissible soil salinity in the balance refers to the field saturation or soil moisture content at field capacity. The EC value of a saturated paste in the laboratory contains about twice as much water. So the ECe of the extract of the paste would be about 8/2 = 4 mmho/cm or dS/m.

Note : If the irrigation requirement for salinity control is calculated only with the data of the irrigation season, one obtains:

$I_g = 457$ mm, instead of 800 mm.

Using $I_g = 457$ mm, and taking the water balance of the non- irrigation season into account, this would result in a salt concentration C_r greater than the permissible salt concentration C_{rp} assumed at EC = 8 mmho/cm or dS/m. In fact, when $I_g < 700$ mm (the total evaporation minus the rainfall), there can be no equilibrium value of C_r, but it will increase continuously.

It is concluded, therefore, that salt balance calculations should be made for longer periods of time, instead of for seasons only, especially when the upward seepage of ground water and the seasonal storage plays an important role.

CHAPTER - 13

GLOSSARY

Alkali Soils: Alkali soils are characterised by an exchangeable sodium per centage (ESP) >15 and pH of the saturated soil paste between 8.5 and 10. The electrical conductivity of these soils normally is less than 4 dS/m.

Anion: Negatively charged constituent or ion in the water. Chloride, sulfate, and bicarbonate are anions.

Attainable leaching fraction: The smallest average leaching fraction required under a given set of conditions to satisfy crop needs and control salinity in the least-watered parts of the field.

Bio-drainage: Bio-drainage is lowering the water table through bio-energy of the trees like eucalyptus.

Capacity of tile drain: The discharge through the tile drains can be calculated either by using the Hooghoudt's approach for steady state conditions or the USBR's approach for the unsteady state condition.

Cation exchange capacity: Relative capacity of positively charged ions (cations) attached to clay particles in a given soil to be exchanged for other types of cations in the soil solution. Too much sodium on the clay particles relative to calcium and magnesium can cause the clay to swell, making the soil less permeable to water.

Cation: Positively charged constituent or ion in the water. Sodium, calcium, magnesium, and potassium are cations.

Chlorosis: Yellowing or bleaching of leaves, often induced by a nutrient deficiency, specific-ion toxicity, or disease.

Composite drainage system: It is a drainage system in which all field drains and all collector drains are buried pipes.

Continuous ponding: The process of reclaiming saline soils by ponding water on the soil surface until enough salt has been removed from the crop root zone.

Crop water use: The amount of water used by a specific crop in a given period of time.

Deep percolation: The phenomenon of irrigation water flowing through the soil past the root zone where it is lost to crop production.

Distribution uniformity (DU): A measure of how uniformly water is applied over a field, calculated as the minimum depth of applied water, divided by the average depth of applied water, multiplied by 100.

Ditch conduit condition: Ditch conduit conditions occur in narrow ditches when conduits are laid underground. No differential settlement of soil is noted in this case.

Drainage requirement: It is the amount of water that must be removed from an area within a certain period so as to avoid an unacceptable rise in the levels of the groundwater or surface water.

Drainage: It is the removal of excess water and dissolved salts from the surface and subsurface of the land in order to enhance crop growth.

Electrical conductivity: The extent to which water conducts electricity, which is proportional to the concentration of dissolved salts present and is therefore, used as an estimate of the total dissolved salts in soil water. Electrical conductivity is expre-ssed in millimhos per centimeter (mmhos/cm) or decisiemens per meter (dS/m): ECi, ECiw, or ECw = electrical conductivity of the irrigation water; ECsw = electrical conductivity of the soil water; ECe = electrical conductivity of the saturated soil extract.

Electrical conductivity (EC): It is a measure of the concentration of salts, defined as the conductance of a cubic centimetre of water at a standard temperature of 250 C.

Evapotranspiration: The amount of water used by a specific crop in a given period of time, comprised of water evaporating from the soil and water transpiring from the plants. Crop evapotranspiration estimates are available from the California Department of Water Resources CIMIS program and from University of California Cooperative Extension offices as either historical averages or real-time estimates.

Exchangeable Sodium Per centage (ESP): The per centage of exchangeable sodium that occupies the total cation exchange capacity of the soil. ESP can be calculated from the following formula: *ESP = Exchangeable (meq/100g) / Cation exchange capacity (meq/100g) x 100.*

Field drainage system: It is a network that gathers the excess water from the land by means of field drains, possibly supplemented by measures to promote the flow of water to these drains.

Foliar absorption rate: Rate at which constituents in water are absorbed by plant leaves. Glycophytes. A group of plants adversely affected by salinity. Most crop plants are glycophytes. Halophytes. Plant group capable of tolerating relatively high levels of salinity.

Glover formula: The Equation is used to find out the drain spacing in unsteady state conditions for sub-surface.

Grade of tile drains: Tile drains grades should be such that no sedimentation occurs in the tiles.

Hooghoudt's Equation: This equation is used to find out the depth and spacing in steady state condition for sub-surface and surface drains.

Hydraulic conductivity: It is a measure of the water-transmitting capacity of a soil.

Hydraulic conductivity: The ease with which water flows through the soil, determined by the physical pro- perties and water content of the soil.

Infiltration rate: The rate at which water infiltrates the soil, usually expressed in inches or centimeters per hour.

Interceptor drain: Usually a single drain line installed perpendicular to the direction of groundwater flow, used to remove shallow groundwater flowing from upper-lying areas or to intercept seepage from waterways.

Intermittent ponding: A method of reclaiming saline soil by ponding small amounts of water on the soil surface in a wetting and drying cycle.

Ion: A positively or negatively charged constituent in water. Cations are positively charged ions and anions are negatively charged ions. Sodium, calcium, magnesium, and potassium are cations, and chloride, sulfate, and bicarbonate are anions.

Irrigation efficiency: A measure of the portion of total applied irrigation water beneficially used – as for crop water needs, frost protection, salt leaching, and chemical application – over the course of a season. Calculated as beneficially used water divided by total water applied, multiplied by 100.

Land grading: It is forming the surface of the land to predetermined grades, so that each row or surface slopes to a field drain.

Land planning: It is smoothing the land surface with a land plane to eliminate minor depressions and irregularities without changing the general topography.

Leaching fraction: The fraction of infiltrated water applied beyond the soil moisture depletion level, which percolates below the root zone as excess water.

Leaching Requirement: It is defined as the ratio of equivalent depth of drainage water to the depth of irrigation water expressed as a fraction or per cent.

Leaching requirement: The leaching fraction needed to keep the root zone salinity level at or below the threshold tolerated by the crop. The leaching fraction is determined by the crop's tolerance to salinity and by the salinity of the irrigation water.

Leaching: Applying irrigation water in excess of the soil moisture depletion level to remove salts from the root zone.

Leaching: It is the removal of soluble salts by water percolating through the soil.

Leaching: Leaching is the process of dissolving and transporting soluble salts by downward movement of water through the soil.

Load on drain pipe: It is the dead load of the soil on the pipe.

Main drainage system: It is a water-conveyance system that receives water from the field drainage systems, surface runoff and groundwater flow, and transports it to the outlet point.

Man-made drainage system: It is an artificial system of surface drains and/or subsurface drains, related structures, and pumps (if any) to remove excess water from an area.

Necrosis: Plant condition indicated by the presence of dead tissue, often induced by an extreme nutrient deficiency, disease, or specific ion toxicity.

Outlet: It is the terminal point of the entire drainage system, from where the drainage water is discharged into a river, a lake, or a sea.

Parallel drainage system: Drainage system consisting of buried perforated pipe placed at equal intervals throughout a field for draining away subsurface water caused by deep percolation through the overlying land. Also called a relief drainage system.

Piezometer: Device for monitoring groundwater depth and movement by measuring the hydraulic head at a point below the water table or water level.

Polymers: Soil amendments reputed by manufacturers to react with lime in the soil to supply free calcium.

Ponding: It is the accumulation of excess water on the soil surface.

Pre-irrigation reclamation method: A method of estimating the amount of irrigation water needed for leaching to reduce soil salinity to acceptable levels during pre-irrigations.

Projecting conduit conditions: These conditions occur in underground conduits laid in wider trenches which leads to differential settlement of soil prisms when the conduit rests on undisturbed soil.

Saline Soils: Saline soils are characterised by pH < 8.2, ESP < 15 and EC > 4 dS/m.

Saline/sodic soil: Soil affected by both excess salt and excess sodium.

Saline-Alkali Soils: There soils have character of both saline and alkali soils and have EC > 4 dS/m, ESP >15 and pH about 8.5.

Salinity: Soil condition in which the salt concentration in the crop root zone is too high for optimum plant growth and yield.

Salinization: It is the accumulation of soluble salts at the soil surface, or at some point below the soil surface, to levels that have negative effects on plant growth and/or on soils.

Singular drainage system: It is a drainage system in which the field drains are buried pipes and all field drains discharge into open collector drains.

Size of tile drain: The size of the tile drains is determined using the maximum expected flow and the grade. Manning's formula is used for the design.

Sodicity: Condition in which the salt composition of the soil within the crop root zone is dominated by sodium, which affects soil structure and water infiltration.

Sodium adsorption ratio (SAR): Relationship between the concentration of sodium (Na) in the irrigation water relative to the concentrations of calcium (Ca) and magnesium (Mg), expressed in meq/l as follows:

$$SAR = \frac{Na}{\sqrt{\frac{Ca + Mg}{2}}}$$

Soil amendment: A substance added to the soil primarily to improve its physical condition.

Specific-ion toxicity: Injury to the plant caused by a specific constituent, usually chloride, boron, or sodium, that has accumulated in a particular part of the plant, such as leaves and stems.

Steady state: For conditions where the rainfall rate or irrigation, rate is constant, the drain discharge will also be constant if the water table does not change with time.

Subsurface drainage system: It is a man-made system that induces excess water and dissolved salts to flow through the soil to pipes or open drains, from where it can be evacuated

Subsurface drainage: It is the removal of excess water and dissolved salts from soils via groundwater flow to the drains, so that the water table and root-zone salinity are controlled.

Surface drainage system: It is a system of drainage measures, such as open drains and land forming, to prevent ponding by diverting excess surface water to a collector drain.

Surface drainage: It is the removal of excess water from the surface of the land by diverting it into improved natural or constructed drains, supplemented, when necessary, by the shaping and grading of the land surface towards such drains.

Tile strength: Maximum bearing capacity of the load on concrete or clay tiles. It is determined by three-edge bearing test and sand bearing test.

Total dissolved solids (TDS): A measure of the dissolved solids in soil water, expressed in either parts per million or milligrams per liter, used to estimate the relative salinity hazard of the water.

Water amendment: Chemicals added to water to improve soil-water properties, such as water infiltration.

Water Table: It is the locus of points at which the pressure in the groundwater is equal to atmospheric pressure.

Waterlogging: It is the accumulation of excess water in the root zone of the soil.

REFERENCES

Anonymous. 1972. Handbook of Hydrology. Central Unit for Soil Conservation (Hydrology and Sedimentation). Soil Conservation Division. Ministry of Agriculture, Government of India, New Delhi. pp. 2-11.

Aronovici, V.S. 1947. The mechanical analysis as an index of subsoil permeability. Soil Science Society of American Proc. 11, pp. 137-141.

Ayers, R.S. and D.W. Westcott. 1985. Water quality for agriculture. Rev. ed. Irrigation and Drainage Paper 29, FAO, Rome, 174 p.

Bentley, W.J., R.W. Skaggs and J.E. Parsons 1989. The effect of variation in hydraulic conductivity on water table drawdown. Technical Bulletin, North Carolina Agricultural Research Service, North Carolina State University, 288, Raleigh, 23 p.

Bouma, J., A. Jongerius, and D. Schoonderbeek 1979. Calculation of saturated hydraulic conductivity of some pedal clay soils using micromorphometric Data. Soil Science Society of American Journal 43, pp. 261-264.

Boumans, J.H. 1976. Drainage calculations in stratified soils using the anisotropic soil model to simulate hydraulic conductivity conditions. In: J. Wesseling (ed.), Proceedings of the international drainage workshop. ILRI Publication 25, Wageningen, pp. 108-123.

Bouwer, H., and R.D. Jackson 1974. Determining soil properties. In: J. van Schilfgaarde (ed.), Drainage for Agriculture. Agronomy 17. American Society of Agronomy, Madison, pp. 611-672.

Camp, C.R. 1977. Determination of hydraulic conductivity for a Louisiana alluvial soil. Third National. Drainage Symposium Proceedings. American Society Agricultural Engineers, Michigan, pp. 1-77.

CWC (Central Water Commission). 1997. Irrigated area drainage problems in India. In: Proceedings of the ICID seventh International Drainage Workshop "Drainage for the 21st Century", Vol., 1 pp. C5/1-C5/2; Malacca, Malaysia

Dieleman, P.J. and B.D. Trafford 1976. Drainage testing. Irrigation and Drainage Paper 28, FAO, Rome, 172 p.

Dumm, Lee D. 1954. New formula for determining depth and spacing of subsurface drains in irrigated lands. Agric. Engineering. 35. pp. 726-30.

FAO. 1985b. Water quality for agriculture, by R.S. Ayers & D.W. Westcot. Irrigation and Drainage Paper No. 29 (Rev. 1). Rome.

FAO. 1999. India. In: Irrigation in Asia in Figures, Water Reports 18, pp. 91-106, Food and Agricultural Organization of the United Nations, Rome, Italy.

Fipps, F. and R.W. Skaggs. 1989. Influence of slope on subsurface drainage of hillsides. Water Resour. Res. 25(7). 1717-1726.

Guyer, P.Q. 1996. Livestock water quality. G79-467-A. Cooperative Extension, Institute of Agriculture and Natural Environment, University of Nebraska-Lincoln, The United States of America. http://www.ianr.unl.edu/pubs/beef/g467.htm

Hoffman, G.J. 1986. Guidelines for reclamation of salt-affected soils. Appl. Agric. Res. 1(2): 65-72.

Hooghoudt, S.B. 1940. General consideration of the problem of field drainage by parallel drains, ditches, watercourses, and channels. Publ. No. 7 in the series Contribution to the knowledge of some physical parameters of the soil (titles translated from Dutch). Bodemkundig Instituut, Groningen, The Netherlands.

Kitamura, Y.; K. Murashima; and Y. Ogino 1997. "Drainage in Asia (2): Manifold Drainage Problems and their Remedial Measures in India" In: Rural and environmental engineering, No. 32 (1997.2) pp. 22-41, Tokyo, Japan.

LeSaffre, B. 1987. Analytical formulae for traverse drainage of sloping lands with constant rainfall. Irrig. Drain. Syst. I. 105-121.

Maas, E.V. & Grattan, S.R. 1999. Crop yields as affected by salinity. In R.W. Skaggs & J. Van Schilfgaarde, eds. Agricultural drainage. Number 38 in the series Agronomy. Madinson, Wisconsin, The United States of America, American Society of Agronomy, Crop Science Society of America, Soil Science Society of America.

Marshall, T.J. 1957. Permeability and the size distribution of pores. Nature 180, pp. 664-665.

Oosterbaan, R.J. 1990a. Single pipe drains with entrance resistance above a semi-confined aquifer. Symposium on Land Drainage for Salinity Control in Arid and Semi-Arid Regions, Vol. 3, 36-46, Cairo, Egypt.

Oosterbaan, R.J. 1990b. Parallel pipe drains above a semi-confined aquifer with upward seepage. Symposium on Land Drainage for Salinity Control in Arid and Semi-Arid Regions, Vol. 3, 26-35, Egypt.

Oosterbaan, R.J., A. Pissarra and J.G. van Alphen 1989. Hydraulic head and discharge relations of pipe drainage systems with entrance resistance. Proceedings 15th European Conference on Agricultural Water Management Vol. III: Installation and Maintenance of Drainage and Irrigation Systems. 86-98. ICID, Dubrovnik, Yugoslavia.

Oosterbaan, R.J. 1975. Interception drainage and drainage of sloping lands. Bull. of the Irrig. Drain. and Flood Contr. Res. Council of Pakistan (5)1. 1-16.

Oosterbaan, R.J. 1986. Tubewell-spacing formulas for subsurface drainage. In: K.V.H. Smith and D.W. Rycroft (Eds.), Hydraulic Design in Water Resources Engineering: Land Drainage. 75-84. Proceedings of the 2nd International Conference, Southhampton University, U.K.

Oosterbaan, R.J., D.P. Sharma, K.N. Singh and Rao K.V.G.K. 1990. Crop production and soil salinity: evaluation of field data from India by segmented linear regression. In: Symposium on land drainage for salinity control in arid and semi-arid regions, Vol. III. Drainage Research Institute, Cairo. pp. 373-383.

Rao, K.V.G.K., Kumbhare, P.S., Kamra S.K. and Oosterbaan R.J. 1990. Reclamation of waterlogged saline aluvial soils in India by subsurface drainage. In: Symposium on land drainage for salinity control in arid and semi-arid regions, Vol. 2. Drainage Research Institute, Cairo. pp. 17-25.

Rao, K.V.G.K., Sharma D.P. and Osterbaan R.J. 1992. Subirrigation by groundwater management with controlled subsurface drainage in semi-arid areas. International Conference on Supplementary Irrigation and Drought Management, Bari, Italy

Reynolds, W.D. and D.E. Elrick 1985. In-situ measurement of field saturated hydraulic conductivity, sorptivity, and the a-parameter, using the Guelph permeameter. Soil Science, 140, 4, pp. 292-302.

Rudd, A.V. and Chardon C.W. 1977. The effects of drainage on cane yields as measured by water table height in the Machnade Mill area. In: Proceedings of the 44th Conference of the Queensland Society of Sugar Cane Technology, Australia.

Schultz, B. (ed.). 1990. Guidelines on the Construction of Horizontal Subsurface Drainage Systems. Working Group on Drainage Construction, ICID, New Delhi.

Schwab, G.O., Taylor, G.S., Fouss J.L. and Ehud Stibble 1966. Crop response from tile and surface drainage. Soil Science Society American Proceedings, 30, pp. 634-637.

Sharma, D.P., Singh, K.N. and Kumbhare, P.S. 2000. Sustainable use of saline drainage water for irrigation in semi-arid regions. In Proc. of the 8th ICID International Drainage Workshop, "Role of drainage and challenges in 21st century". Vigyan Bhawan, New Delhi. Vol. 3: 215-226.

Sing, O.P. and K.K. Datta. 1997. Problems, research and adopting drainage technology for reclamation of waterlogged and saline soils in India. In: Proceedings of the ICID seventh International Drainage Workshop "Drainage for the 21st Century", Vol. 3, Malacca, Malaysia.

Smedema, L.K. and D.W. Rycroft 1983. Land drainage : planning and design of Agricultural Drainage Systems. Batsford, London, 376 p.

Smith, R.E. and Goodrich, D.C. (2005). Rainfall Excess Overland Flow. Encyclopedia of hydrological sciences. pp.1707-1718.

U.S. Army Corps of Engineers. 1941. Investigation of filter requirements for underdrains. Eng. Water ways Exp. Sta. Tech. Memo. Pp. 183-4.

USBR, 1978. Drainage manual. United States Department of Interior, Bureau of Reclamation, Denver, CO. Unites States.

USBR. 1993. Drainage manual. 2nd Ed. United States Department of Interior, Bureau of Reclamation, Denver, CO. Unites States.

USSCS. 1973. Drainage of Agricultural Lands. A practical handbook for the planning, design, construction and maintenance of agricultural drainage systems. USDA Soil Conservation Service. Water Information Center lac. Washington, NY, United States.

Van Beers, W.F.J. 1970. The auger-hole method : a field measurement of the hydraulic conductivity of soil below the water table. Rev. ed. ILRI Bulletin 1, Wageningen, 32 p.

Van Schilfgaarde, J., 1963. Design of tile drainage for falling water tables. Jour. Irrig. and Drain. Div., ASCE 89, pp. 1-11.

Wesseling, J. 1973. Subsurface flow into drains. Drainage Principles and Applications Vol. II: Theories of Field Drainage and Watershed Runoff. 2-56. Publ. 16. ILRI, Wageningen, The Netherlands.

WHO, 1996. Guidelines for drinking-water quality, 2nd ed. Vol. 2 Health criteria and other supporting information, p. 940-949; and WHO, 1998. Addendum to Vol. 2. p. 281-283. Geneva, World Health Organization. Summary tables. http://www.who.int/water_sanitation_health/GDWQ/Summary_tables/Sumtab.htm

Wit, K.E. 1967. Apparatus for measuring hydraulic conductivity of undisturbed soil samples. Technical Bulletin 52. Institute for Land and Water Management Research, Wageningen, 12 p.

World Bank. 1998. India – Water resources management sector review, Report on the Irrigation Sector, Washington D.C., USA.

Wösten, J.H.M. 1990. Use of soil survey data to improve simulation of water movement in soils. Thesis, Agricultural University, Wageningen, 103 p.

Zangar, C.N. 1953. Theory and problems of water percolation. U.S. Bureau of Reclamation. Engineering Monograph No. 8, Denver, 76 p.